IMMUNOLOGY OF BACTERIAL POLYSACCHARIDES

Proceedings of a Symposium presented at the Annual Meeting of the American Society for Microbiology, Las Vegas, Nevada, 16 May 1978

Editors:

JON A. RUDBACH
Abbott Laboratories, North Chicago, Illinois; Department of Microbiology, University of Montana, Missoula, Montana, U.S.A.

PHILLIP J. BAKER
Laboratory of Microbial Immunity, National Institute of Allergy and Infectious Diseases, National Institutes of Health, Bethesda, Maryland, U.S.A.

ELSEVIER/NORTH-HOLLAND
NEW YORK • AMSTERDAM • OXFORD

Published by:

Elsevier North Holland, Inc.
52 Vanderbilt Avenue, New York, New York 10017

Sole distributors outside of the U. S. A. and Canada:

Elsevier/North-Holland Biomedical Press
335 Jan van Galenstraat, P. O. Box 211
Amsterdam, The Netherlands

Library of Congress Cataloging in Publication Data

Symposium on immunology of bacterial polysaccharides,
Las Vegas, Nev., 1978.
Immunology of bacterial polysaccharides.
(Developments in immunology; v. 2 ISSN 0163-5921)

Bibliography: p.
Includes index.
1. Microbial polysaccharides — Congress. 2. Antigens — Congress. 3. Antigen-antibody reactions — Congress. 4. Immunology — Congress. I. Rudbach, Jon A. II. Baker, Phillip III. Title. IV. Title: Bacterial polysaccharides V. Series.
QR92P6S95 1978 616.01'4 78-31961
ISBN 0-444-00315-0

Manufactured in the United States of America

Contents

Contributors

PHILLIP J. BAKER, Laboratory of Microbial Immunity, National Institute of Allergy and Infectious Diseases, National Institutes of Health, Bethesda, Maryland, 20014, USA

JENN C. CHEN, Department of Immunology and Microbiology, Wayne State University Medical School, Detriot, Michigan, 48201, USA

RONALD GOLD, Department of Pediatrics, University of Connecticut School of Medicine, Farmington, Connecticut, 06032, USA

THOMAS J. KINDT, Laboratory of Immunogenetics, National Institute of Allergy and Infectious Diseases, National Institutes of Health, Bethesda, Maryland, 20014, USA

TUAN-HUEY KUO, Department of Immunology and Microbiology, Wayne State University Medical School, Detroit, Michigan, 48201, USA

MYRON LEON, Department of Immunology and Microbiology, Wayne State University Medical School, Detroit, Michigan, 48201, USA

BENJAMIN PRESCOTT, Laboratory of Microbial Immunity, National Institute of Allergy and Infectious Diseases, National Institutes of Health, Bethesda, Maryland, 20014, USA

JON A. RUDBACH, Abbott Laboratories, North Chicago, Illinois, 60064; Department of Microbiology, University of Montana, Missoula, Montana, 59812, USA

JAMES WATSON, Department of Medical Microbiology, University of California, Irvine, California, 92717, USA

MARTIN L. YARMUSH, Laboratory of Immunogenetics, National Institute of Allergy and Infectious Diseases, National Institutes of Health, Bethesda, Maryland, 20014, USA

Preface

This book contains the proceedings of a symposium on the Immunology of Bacterial Polysaccharides that was held in Las Vegas, Nevada on May 16, 1978. When it was proposed by Natalie Cremer, in 1977, that we organize a Divisional Group Symposium for the 1978 annual meeting of the American Society for Microbiology, two possibilities occurred to us. First, we could plan a strong program of immunology for presentation at the annual meeting which would re-emphasize that the science of immunology was a "child" of microbiology, a fact that has been somewhat forgotten or ignored during the past decade. Second, through the vehicle of the symposium, the contributions that studies with bacterial polysaccharides have made - and are now making - to the overall field of immunology and immunochemistry could be highlighted. We elected to take the latter approach since many of the contributions considered deal with issues that are of great interest, not only to those currently involved in basic research in immunology, but also to those wishing to apply the knowledge gained to the development of more effective and improved vaccines against infectious agents.

It was a difficult task to choose the limited number of individuals who could be invited to contribute to the symposium. Admittedly, the Heidelbergers, Kabats, Staubs, Kauffmans, Lüderitzes, Westphals, Lederers, Springers, and many others are the exemplars in studies on bacterial polysaccharide antigens; indeed, their accomplishments are well-known and it is to them and their colleagues that this book is dedicated. However, after much soul-searching and thought, we decided to select a group of mid-career investigators who would present examples of the types of studies now being pursued in this area of research and who would indicate the new directions in which work in this field is moving. It will be noted that the list of contributors is short, mainly because of the amount of time alloted for the symposium (one morning session), and that all of the contributors are from the United States. The latter choice was deliberate on our part, not because we wished to be prejudicial or to minimize the importance of work being done by others throughout the world, but simply because funds were not available to defray the costs of invited participants; the expense of foreign travel was too much to ask our colleagues to bear. We deeply regreted the fact that the contributions of many other outstanding investigators could not be included in this symposium, which was much too brief to encompass a topic of this scope. However, it is our hope that this initial symposium will stimulate a response, or even an appeal, for a much more expanded and broader symposium on this topic in the future; we invite the comments and suggestions of all interested parties in this regard.

This is an appropriate place for us to thank, again, the participants in the symposium, all of whom contributed to this volume. They did so without thought of recompense. The final draft of the manuscript in camera-ready copy was typed by Robin Ziesemer and Phyllis McCullough of Abbott Laboratories. We owe a debt of gratitude to them and their organization. Finally, we wish to acknowledge the encouragement and aid of Yale Altman and Susan Koscielniak of Elsevier-North Holland for getting these presentations from the symposium into book form.

JON A. RUDBACH

PHILLIP J. BAKER

October, 1978

Rudbach/Baker, eds. Immunology of Bacterial Polysaccharides

CONTRIBUTIONS OF STUDIES WITH BACTERIAL POLYSACCHARIDE ANTIGENS

JON A. RUDBACH* AND PHILLIP J. BAKER**
*Abbott Laboratories, North Chicago, Illinois 60064 and Department of Microbiology, University of Montana, Missoula, Montana 59812; **Laboratory of Microbial Immunity, National Institute of Allergy and Infectious Diseases, National Institutes of Health, Bethesda, Maryland 20014

INTRODUCTION

Interest in the use of bacterial polysaccharides as experimental models in studies on basic immunological and immunochemical processes has waxed and waned over the years. In the hay-days of the logarithmic growth of modern experimental immunology during part of the 1950s and 1960s, the "in-group" of immunological trend-setters considered investigations with microbial polysaccharide antigens as crude or weird. The fad dictated that most relevant data were obtained from investigations using crystalline serum or egg protein antigens. This philosophy was in direct contrast with the previous generation of work, wherein studies with bacterial polysaccharides provided the first estimate of the size of the antibody combining site, the first experimental example of specific immunological unresponsiveness, and the first glimpse of the structures of immunodeterminant groups in natural products.

Beginning in the late 1960s and accelerating rapidly during the first half of the 1970s, interest in immunological studies with bacterial polysaccharides received a new burst of energy. The reasons for this can be traced to several key discoveries. The first of these was the observation that certain classes of antigens, most notably polysaccharides, could stimulate the generation of humoral antibody responses in the absence of thymus-derived (T) lymphocytes. This resulted in the concept of classifying antigens as being either T cell-independent or T cell-dependent. A second key observation was that bacterial lipopolysaccharide antigens could activate complement by a mechanism that circumvented the necessity for antibody, thus bypassing the early-acting components. A third key development was the demonstration that monoclonal, myeloma proteins could react, in a manner similar to antibodies, with natural polysaccharide antigens, and that, in certain cases, hyperimmunization of animals with bacterial polysaccharides would result in monoclonal, or at least in highly restricted types of antibody responses.

The following is not intended to be a complete survey of the literature on bacterial polysaccharide antigens, nor is it representative of all types of studies conducted with these materials. Instead, a few examples are given of the contributions that studies with bacterial polysaccharide antigens have made to immunology and to serology. Inasmuch as this survey is brief and incomplete, direct citations, and the implied credit-of-discovery, are not made in the text. Instead, an alphabetical list of general references to the literature is provided.

BACTERIAL POLYSACCHARIDES AS BASES FOR SEROLOGICAL GROUPING OF MICROORGANISMS

From the 1920s into the 1950s, Lancefield and her collaborators established a system for classifying or grouping streptococci based on the presence of serologically different polysaccharides in the bacterial cell wall. Each group was shown to possess its own unique polysaccharide antigen. Furthermore, this grouping system was found to be more than a toy for the serotaxonomist, because a correlation could be established between the presence of certain antigenic groups and disease. For example, the majority of Streptococcus pyogenes isolated from severe lesions in humans belonged to Group A, whereas those from cattle with mastitis belonged to Group B, and so forth.

The existance of antigenically different types of pneumococci was first noted by Neufeld and Händel. Shortly thereafter, Dochez, Avery, and coworkers initiated the study of antigenic relationships among a large group of pneumococcal isolates. From this work, a pattern emerged that resulted in the grouping of Streptococcus pneumoniae strains into three antigenically distinct types. The serologically active materials were extracted from the pneumococci by Avery, Heidelberger, and coworkers in the 1920s. It was found that the type-specific materials were complex polysaccharides that came from the pneumococcal capsule. From a serological point of view, the capsular polysaccharides of S. pneumoniae appeared to be analogous to the group specific polysaccharides of the Lancefield schema for S. pyogenes. Clinical relevance of the pneumococcal capsular polysaccharide was obtained through the discovery that immunity to pneumococcal disease was specific for the type of specific capsular polysaccharide in question.

Probably the most ambitious bacterial seroclassification

ever attempted was the Kauffmann-White scheme. Separate serotypes have been described for over 1000 salmonellae. The bases for classification were immunodeterminant groups on the O-somatic, or cell wall, antigen of the bacteria. O-antigenic determinants were found in the polysaccharide portion of the lipopolysaccharide of enterobacteria. Similar serotyping schemes, based on the O-immunodeterminants have been established for *Escherichia coli* and *Shigella species*.

With *Neisseria meningitidis* another example is found of a bacterial classification system based on serological reactivity of a polysaccharide antigen. Meningococci were divided into nine serological groups on the basis of the antigenic reactivity of their polysaccharide capsules. The immunological specificity of these serogroups, again, has more than taxonomic importance, because antibodies against the various group specific polysaccharides have been shown to confer specific protection against infection in man.

This brief summary of a few classification schemes based on the immunological specificity of bacterial polysaccharides was not intended to be exhaustive. Other clinically important bacteria which have specific polysaccharide antigens that can be used as a basis for classification include *Klebsiella species*, the cholera vibrio, and various rickettsia and fungi.

THERAPEUTIC USE OF PASSIVE ANTISERUM TO PNEMOCOCCAL POLYSACCHARIDE

Following the discovery of the different antigenic types of complex capsular polysaccharides on pneumococci, it was shown that antibodies to these capsular antigens protected experimental animals against pneumococcal disease. Furthermore, humans recovering "by crisis" from lobar pneumonia, did so because of the appearance of serum antibodies specific for the pneumococcal capsule. A logical extension of these observations was the use of passively administered antipneumococcal serum for the cure of, as well as for protection against, pneumonia in man. It was found that the dose of passively administered antiserum that would cure patients was manyfold higher than the dose which resulted in protection; up to 500 ml of horse antipneumococcal serum was given intravenously, and such a therapeutic dose reduced the mortality due to type I pneumococcal pneumonia by one-half of that expected. The effectiveness of such serum therapy was increased and side reactions were reduced when the antibody-containing euglobulin fraction was used, instead of whole serum. During the 1930s, commercial sera were produced for over 75 specific types of pneumococcal capsules. These were first produced in horses and then later in rabbits. However, the use of passively administered antisera for the treatment of pneumococcal pneumonia was discontinued with the advent of the chemotherapeutic-antibiotic era. Although, the logistic problem of having to treat each specific type of pneumococcus with its own specific antisera speeded the abandonment of this type of therapy, it is now known that only about 12 different types of specific pneumococci account for more than 85% of confirmed cases of pneumococcal infections in man.

DEVELOPMENT OF THE QUANTITATIVE PRECIPITIN REACTION

The birth of immunochemistry as a precise science probably

was coincidental with the development of the quantitive precipitin reaction by Heidelberger and his colleagues. This technique was first demonstrated with a pneumococcal capsular polysaccharide-antipolysaccharide system. The washed precipitates were analyzed quantitatively for nitrogen, and since the polysaccharide antigen contained no nitrogen, the nitrogen values of the precipitates could be converted directly into antibody protein. For the first time, the absolute mass of specific antibody in antiserum could be determined precisely. This task could not have been accomplished as easily without the judicious use of a highly purified bacterial polysaccharide antigen.

SIZE OF THE ANTIBODY COMBINING SITE

A basic question in immunology concerned the size of the antibody-combining site. Direct measurements of antibody combining sites are technically very difficult, if not almost impossible to obtain, even with our modern level of technology and instrumentation. Elvin Kabat, however, solved this problem in a simple, logical manner; he said that the antibody combining site was as large as the immunodeterminant group that filled it. Also, since short range forces played a major role in holding the immunodeterminant group in the combining site, the shape of the site must conform quite closely to the shape of the immunodeterminant group.

Kabat demonstrated, with a dextran-antidextran system in which quantitative precipitation was inhibited by the addition of oligosaccharides of known sizes, that the maximum size of the antibody combining site was about six glucose units. This type of investigation was possible because terminal nonreducing glucose residues form the immunodeterminant groups of dextran. Furthermore, it was found that each successive sugar residue contributed more and more to the tightness of binding in the antibody site, until a maximum was reached. Thus, a quantitative estimate of the fractional "fill" of the antibody sites by the oligosaccharides could be obtained. It turned out that a molecule of isomaltohexose with dimensions of 3.4 x 1.2 x 0.7 nm would just fill the antibody combining site.

ELUCIDATION OF IMMUNODETERMINANT GROUPS OF (AND PRECISE ANTIBODY SPECIFICITY FOR) NATURAL PRODUCTS

The meticulous and classic studies of Karl Landsteiner on the serological reactions of artificial antigens, prepared by complexing simple chemicals to protein carriers, formed the basis for our understanding of the specificity of serological reactions. Landsteiner also investigated the basis of serological specificity of several natural antigens. In general, difficulties were encountered in attempts to split proteins into immunodeterminant groups that retained serological specificity. One of the exceptions was a fibrous protein, silk fibroin, from which a short polypeptide could be obtained that would inhibit antibody from combining with the homologous, or native, antigen.

Bacterial polysaccharide antigens were objects of early investigations that elucidated the nature of immunodetermant groups present in natural products. Serological specificties of capsular polysaccharides from pneumococci were determined by the inhibition of quantitative precipitin reactions by simple sugars and sugar acids. Artificial antigens, prepared from sugar acids, complexed with appropriate carriers, crossreacted with the natural polysaccharides; in fact, immunization with such materials even conferred protection against disease. These procedures provided direct evidence about the chemical composition of immunodeterminant groups. Not so direct, but cogent evidence on the nature of immunological specificity of pneumococcal polysaccharides was provided by the demonstration that selective esterification would diminish immunological reactivity. Overlapping reactions and crossreactions between well-characterized polysaccharide antigens and uncharacterized, naturally occurring substances also provided presumptive evidence for immunodeterminant group structure.

It was shown that the terminal non-reducing ends of polysaccharide chains or side chains, in many cases were the major, if not sole, immunologically reactive sites on polysaccharides. This observation paved the way for isolation of the immunodeterminant groups from hydrolyzed polysaccharides; also, synthesis of artificial immunodeterminant groups could be attempted, if a presumptive structure had been established.

Some of the most complete and sophisticated work on the structure of immunodeterminant groups on polysaccharide antigens emerged from investigations on the serological specificity of the O-antigens in lipopolysaccharides of enterobacteria. Staub, Lüderitz, Westphal and others have determined the chemotypes of hundreds of salmonella and *E. coli* strains. During the course of these investigations, several new classes of sugars and sugar acids were discovered and some of them were found to play an immunodominant role in the immunodeterminant groups. The results derived from this chemotyping system have established a structural basis for serotyping schemes, such as the Kauffmann-White system discussed previously.

POLYSACCHARIDE ANTIGENS AND HUMAN BLOOD GROUPS

It is well known that the major human blood group alloantigens are glycoproteins with carbohydrate immunodeterminant groups. If an individual does not possess a particular immunodeterminant sugar (such as α -D-galactose) on the blood group-backbone polysaccharide chain, the blood group B immunodeterminant will be lacking from the erythrocytes of that individual. In this case, isohemagglutinins with Group B specificity will be formed. It has been found that the so-called normal isohemagglutinins to blood group substances are antibodies that have been induced to polysaccharide antigens present on bacteria, other microbes, or substances in the environment. It was in the course of examining cross reactions and structural similarities between bacterial polysaccharide antigens and human blood group antigens that the mechanism for induction of normal isoagglutinins was established.

BACTERIAL POLYSACCHARIDES AS MODELS OF T CELL-INDEPENDENT ANTIGENS

It has been recognized since the 1960s that the cooperative interactions of several types of lymphoid cells were required to generate an antibody response to some antigens. The elegant studies of Claman, as well as those by others, showed that at least two distinct types of lymphocytes had to be present for an immunological response to be initiated to erythrocytic and to certain protein antigens. One type of lymphocyte was derived by division and transformation of reticular cells under the influence of the thymus gland; these are known as thymus-derived or T cells. The other type of lymphocyte matured from bone marrow precursors, without thymic influence and is called a B cell.

As would have been anticipated, such a discovery prompted similar investigations with other antigens. These studies sought to determine if the T cell and B cell requirements described initially were unique phenomena or if the requirements could be applied to all antibody responses, in general. It was shown that both T and B cells were required for generation of antibody responses to a variety of protein and cellular antigens. In addition, it was found that T cells were not required to initiate antibody responses to enterobacterial lipopolysaccharide, to pneumococcal polysaccharide, or to bacterial flagella. Thus, bacterial polysaccharides were recognized as a new class of antigens, the T cell-independent antigens. Much work in the succeeding years made use of the distinctions between T cell-dependent and T cell-independent antigens in efforts to dissect the mechanisms for triggering antibody responses at the cellular and subcellular levels.

PNEUMOCOCCAL POLYSACCHARIDE AND REGULATORY T CELLS

Since the initial observations of systemic adjuvants and immunological tolerance (or paralysis), explanations have been sought for the mechanisms of these phenomena. At least a partial answer was obtained with the discovery of T cell subpopulations that can regulate the magnitude of an antibody response in either a positive or in a negative manner; such regulatory T cells have been referred to as amplifier and suppressor T cells respectively. Regulatory T cells (especially suppressor T cells) may play an important role in the development of autoimmune disease, aging, and immunity to neoplasms, in addition to controlling the systemic humoral antibody response. Continuing investigations on the genetics, prevention, and modification of suppressor and amplifier T cell activities, generated in response to pneumococcal polysaccharide antigens, are yielding information leading to a more complete understanding of the mechanisms of immune regulation in general.

BACTERIAL POLYSACCHARIDE ANTIGENS AND DELAYED HYPERSENSITIVITY

Delayed hypersensitivity was first described in the course of investigations on allergies of infection. Natural extensions of these observations were to fractionate the infectious agents (especially bacteria and fungi) in order to identify the allergen(s). Although the allergenic microorganisms contained copious amounts of antigenic polysaccharides in no cases could naturally developed delayed hypersensitivity be ascribed to the polysaccharide antigens. Moreover, it was found to be extremely difficult, if not impossible in most cases, to elicit a delayed hypersensitivity reaction to polysaccharide antigens in experimental animals. The most likely explanation for the failure of polysaccharide antigens to elicit or to provoke delayed hypersensitivity reactions would be their inability to interact directly with appropriate effector T cells, as evidenced by their T cell-independent nature. However, in the preceeding section it was pointed out that bacterial polysaccharides do stimulate certain subsets of regulatory T cells, even though there is not an absolute requirement for T cells to initiate an antibody response. Certainly, comprehensive and probing investigations with both native and modified polysaccharide antigens could help to define the essential molecular mechanisms that determine the qualitative nature of cellular and humoral immunological responses.

MICROBIAL POLYSACCHARIDES AND THE ALTERNATIVE PATHWAY OF COMPLEMENT ACTIVATION

It is now widely recognized that the latter components of the complement cascade can be activated by two pathways. The first of these in order of discovery, termed the classical pathway, is initiated by an antigen-antibody complex. In the second, termed the alternative pathway, the necessity for the antigen-antibody complexes for the activation of early-acting components of complement is bypassed. A variety of polysaccharides and polysaccharide-rich materials of microbial origin, with certain serum cofactors, can trigger the complement sequence at the C3 step.

The carbohydrate-rich yeast extract, called zymosan, was used extensively by Pillemer and his colleagues to define the serum factor, properdin, that played such a crucial role in the alternative pathway. Rekindling of interest in the properdin pathway was made by the discovery that bacterial lipopolysaccharide was a potent activator of the latter components of complement. Subsequently, it was shown that activation of the alternative pathway by lipopolysaccharide was associated with the polysaccharide portion of the complex.

From a biological view point, the "alternative" pathway of complement activation probably has more significance for the host than does the "classical" pathway. Certainly, polysaccharide antigens of microbial origin had, have, and are being used to dissect this important physiological system.

MICROBIAL POLYSACCHARIDES AND ANTIBODIES WITH RESTRICTED HETEROGENEITY

One of the problems that has hampered structural studies on antibody molecules was the heterogeneity of antibodies isolated from antiserum. Therefore, interpretations of the significance of amino acid sequences of the immunoglobulin polypeptide chains were difficult, if not impossible. Furthermore, antibodies or their fragments could not be isolated with sufficient homogeneity so that meaningful x-ray diffraction data could be obtained. Two general classes of observations offered ways to obtain populations of antibodies with homogeneity sufficient for structural investigations. One of these occurred in the middle 1960s and was the observation that some murine myeloma proteins would precipitate the C-polysaccharide of pneumococci. Subsequently, it was determined that the ligand specificity of these myeloma proteins was directed against the phosphocholine determinants in the pneumococcal polysaccharide. Work which followed showed that extracts of several other bacteria and microbes that reacted with such myeloma proteins all contained phosphocholine, and that this was the reactive site.

When a series of myeloma proteins was tested for reactivity with a variety of potential antigens, it was found that several precipitated with galactan (of nonmicrobial origin); some combined with glucans (dextrans), a polyfructan of *Aerobacter sp.*, and with lipopolysaccharides from *Proteus*, *Salmonella*, and *Escherichia species*. These myeloma proteins were sufficiently homogeneous for meaningful amino acid sequence studies and for x-ray diffraction analysis. Of equal importance was the demonstration that they would bind with antibody-like affinity and specificity to antigens. This lent credibility to the extrapolations from these observations to the situations with, and properties of, "true" antibodies.

The second general class of observations on obtaining antibodies with structural homogeneity was as follows. Human antibodies induced to carbohydrate antigens, such as levans and dextrans, had more molecular homogeneity than did antibodies to other classes of antigens. Results of investigations with these "true" human antibodies of restricted heterogeneity allowed for greater generalizations than could be made from the results obtained with ligand-binding human myeloma proteins. Also, it had been found that certain rabbits, which had been hyperimmunized with streptococcal cell wall carbohydrate or with pneumococcal capsular polysaccharide, produced monoclonal-type antibody responses to the polysaccharides. The best results in obtaining antibodies with restricted heterogeneity were obtained in rabbits immunized intravenously with whole streptococci which had carbohydrate as the outer most cell wall component.

In conclusion, it can be stated that bacterial polysaccharide antigens provided the tools to obtain antibodies of human and other animal origins in states of homogeneity sufficient for structural analyses. Certainly, the understanding of mechanisms of antibody function, immunogenetics, phylogeny of the immunological response, and the pathogenesis of immunoglobulin related disease states have been enhanced through the use of these reagents.

EPITOME

The preceeding was not intended to be a comprehensive review of all contributions that investigations with bacterial polysaccharide antigens have made to immunology. Certainly, it is far from complete. However, polysaccharide antigens, both as model systems for dissecting basic mechanisms of immunological responses and as biologically relevant entities in their own right, do deserve a position of prime importance in the field of immunology. Sometimes the uninitiated tend to over-look polysaccharide antigens in the haste of preparing an experiment. Certainly, very few polysaccharide antigens can be purchased commercially in the highly refined and characterized forms in which they should be used. Thus, additional time is added to the length of an experimental series. This may seem unacceptable in these days of instant accomplishment, success, and gratification. However, those of us who are members of the "club" feel that the extra efforts are well worth the ultimate rewards, and we hope that there might be sufficient interest generated herein to recruit some new members to our group.

GENERAL REFERENCES

1. Basten, A. and Howard, J.G. (1973) in Contemporary Topics in Immunobiology, Vol. 2, Davies, A.J.S. and Carter, R.L. eds., Plenum Press, New York, pp. 265-291.

2. Boyd, W.C. (1966) Fundamentals of Immunology, 4th edition, Interscience Publishers, New York, pp. 1-773.

3. Claman, H.N., Chaperon, E.A. and Triplett, R.F. (1966) Proc. Soc. Exp. Biol. Med. 122, 1167-1172.

4. Claman, H.N., Chaperon, E.A. and Triplett, R.F. (1966) J. Immunol. 97, 828-832.

5. Coutinho, A. and Möller, G. (1975) in Advances in Immunology, Vol. 21, Dixon, F.J. and Kunkel, H.G. eds., Academic Press, New York, pp. 113-236.

6. Götze, O. and Müller- Eberhard, H.J. (1976) in Advances in Immunology, Vol. 24, Dixon, F.J. and Kunkel, H.G. eds., Academic Press, New York, pp. 1-35.

7. Kabat, E.A. (1976) Structural Concepts in Immunology and Immunochemistry, 2nd edition, Holt, Rinehart and Winston, New York, pp. 1-547.

8. Krause, R.M. (1970) in Advances in Immunology, Vol. 12, Dixon, F.J., Jr. and Kunkel, H.G. eds., Academic Press, New York, pp. 1-56.

9. Landsteiner, K. (1962) The Specificity of Serological Reactions, Revised Edition, Dover Publications, Inc., New York, pp. 1-330.

10. Manning, J.K., Reed, N.D. and Jutila, J.W. (1972) J. Immunol. 108, 1470-1472.

11. Mitchell, G.F. and Miller, J.F.A.P. (1968) Proc. Nat. Acad. Sci. U.S.A. 59, 296-303.

12. Nisonoff, A., Hopper, J.E. and Spring, S.B. (1975) The Antibody Molecule, Academic Press, New York, pp. 1-542.

13. Osterland, C.K., Miller, E.J., Karakawa, W.W. and Krause, R.M. (1966) J. Exp. Med. 123, 599-614.

14. Parish, H.J. (1965) A History of Immunization, E. & S. Livingstone Ltd., London, pp. 1-356.

15. Pincus, J.H., Jaton, J.-C., Bloch, K.J. and Haber, E. (1970) J. Immunol. 104, 1143-1148.

16. Potter, M. (1977) in Advances in Immunology, Vol. 25, Kunkel, H.G. and Dixon, F.J. eds., Academic Press, New York, pp. 141-211.

17. Raffel, S. (1961) Immunity, 2nd ed., Appleton-Century-Crofts, Inc., New York, pp. 1-646.

18. Talmage, D.W., Radovich, J. and Hemmingsen H. (1970) in Advances in Immunology, Vol. 12, Dixon, F.J., Jr. and Kunkel H.G. eds., Adacemic Press, New York, pp.271-282.

19. Wilson, G.S. and Miles, A.A. (1955) Topley and Wilson's Principles of Bacteriology and Immunity, 4th edition, The Williams & Wilkins Company, Baltimore, pp. 1-2331.

Rudbach/Baker, eds. Immunology of Bacterial Polysaccharides

GENETIC CONTROL OF RESPONSES TO BACTERIAL LIPOPOLYSACCHARIDES IN MICE

JAMES WATSON
Department of Medical Microbiology, University of California, Irvine, Irvine, California 92717

ABSTRACT

Lipopolysaccharides (LPS) isolated from gram-negative bacteria possess many diverse biological properties. The O-polysaccharide is the major antigenic structure, while the lipid A moiety is responsible for adjuvant effects on the immune system as well as a variety of nonimmune cellular responses.

The expression of a genetic locus, termed Lps, found on chromosome 4 in mice, is required for the initiation of many different cellular responses to lipid A. The importance of understanding the Lps locus lies not only in its use as a tool to dissect the control of the immune system, but in its use in a protective capacity against the endotoxic effect of LPS during bacterial infection.

INTRODUCTION

The bacterial cell wall is a structure of considerable importance to higher organisms. The complex antigenic nature of bacterial cell wall constituents is used by the host to induce immune responses that afford specific protection against bacterial infection. However, a number of cell wall structures exert regulatory effects on the immune system of higher organisms. In particular, the lipopolysaccharides[1] and lipoproteins[2] of gram-negative bacteria, and the peptidoglycans[3] common to most prokaryotes, all possess strong adjuvant properties, and exhibit a wide range of biological effects on cells other than lymphocytes. Because of the host-parasite relationship between higher organisms and bacteria, it is important to understand how the immune system reacts to the "foreign" antigenic determinants presented by the bacterial cell wall and how molecules derived from the cell wall influence the expression of higher cell function. The studies discussed here are aimed at establishing the genetic basis of responses to bacterial lipopolysaccharides in mice. They reveal the complexity of a host-parasite relationship.

Bacterial lipopolysaccharides (LPS) possess many diverse biological properties. The immunological responses elicited by LPS have been of interest for a number of reasons. LPS are extremely potent antigens; small doses elicit very large antigenic responses directed against the 0-specific polysaccharide antigens[4,5]. The lipid A moiety of LPS acts as a specific mitogen for bone marrow-derived (B) lymphocytes in mice[6]. This results in the polyclonal expression of antibody synthesis[7]. In addition, lipid A acts as an adjuvant, both for the specific antibody response to the 0-antigens of the polysaccharide moiety in LPS[8], and for immune responses to such soluble antigens as bovine serum albumin[6]. The lipid A moiety of LPS is also capable of initiating a variety of other effects in mammals, ranging from pyrogenic and toxic effects[9], to macrophage activation[10], nonspecific resistance to bacterial infection[11], and increases in serum levels of growth stimulating substances[12]. An extensive list of these effects is summarized elsewhere[13]. The large number of these endotoxic reactions to LPS have made it difficult to analyze the control mechanisms involved. In particular, the diversity of these responses have both protective and destructive influences on the host organism. The genetic analysis described here reveals that the expression of many of these responses to lipid A involves the expression of a common genetic locus in mice.

The genetic basis of host responses to LPS has been determined largely as a result of the finding that the C3H/HeJ mouse strain is resistant to many of the effects of LPS observed in other mouse strains. These include resistance to the mitogenic, polyclonal, and adjuvant effects of LPS[14], as well as to the toxic effects of LPS[15]. The number of altered LPS response characteristics of C3H/HeJ mice continues to expand. A second mouse strain, C57BL/10. Sc.Cr, has recently been found to be nonresponsive to LPS[16]. Cells from the spleen and lymphocytes of these mice fail to respond to LPS either by proliferation or by increased immunoglobulin synthesis[17].

MATERIALS AND METHODS

Mice

C3H/HeJ and C57BL/6J mice were obtained from the Jackson Laboratory, Bar Harbor, Maine. The backcross mice used in this study were obtained from female (C57BL/6J x C3H/HeJ)F_1 x C3H/HeJ mice. Recombinant inbred (RI) strains derived from the C3H/HeJ and C57BL/6J progenitor strains are described elsewhere[18]. The strains used were BxH-2,3,4,5,6,7,8,10,11,12,14, and 19. C57BL/6J mice have been designated $\underline{Lps}^n$, and C3H/HeJ as $\underline{Lps}^d$. C57BL/6By mice carrying the gene for polysyndactyly (Ps) were obtained from Dr. D. Bailey, and are designated as C57BL/6By-Ps. Male C57BL/6By-Ps/+ were mated with C3H/HeJ females and both the male and female Ps/+ progeny were used in backcrosses to C3H/HeJ parents. Mice were six to eight weeks old at the time of assay.

Major Urinary Protein

Urine samples from individual mice were examined by polyacrylamide gel electrophoresis. C3H/HeJ carry the $Mup\text{-}1^{a}$ allele whereas C57BL/6J mice possess the $Mup\text{-}1^{b}$ allele[19].

Mitogenic Responses

Escherichia coli K235 LPS prepared by a phenol water extraction procedure was generously provided by Abbott Laboratories, N. Chicago, Ill. Spleen cells were cultured with 10 μg/ml LPS, and the incorporation of radioactive thymidine was measured after 48 h, as detailed elsewhere[20].

Hypothermal Responses

Mice were kept in an environment at 22°C. Rectal temperatures were recorded by using a microthermometer (°F). Temperatures were determined in individual mice before injection of 50 μg *E. coli* K235 LPS intraperitoneally. After 1 and 2 h, rectal temperatures were recorded. The rectal probe was inserted at a distance of approximately 1 cm, and the temperature recorded 20-30 seconds later.

Colony-Forming Cells

Colonies containing granulocytes and macrophages were grown by using mouse bone marrow cells cultured in a semisolid agar medium. Female C57BL/6J mice were used as the source of bone marrow cells. Femora were removed and the marrow cells flushed from the medullary cavity with Eagle's medium containing 10% fetal calf serum. The semisolid medium was prepared at 45°C and contained 80% Eagle's medium, 15% fetal bovine serum, and 5% trypticase soy broth (Difco Laboratories, Detroit, Mich.) and 0.3% washed agar which had been boiled for 10 minutes to give a solution. 1 ml of this cell mixture was layered into 35 mm x 12mm plastic dishes (Falcon Plastics, Division of BioQuest, Oxnard, Calif.) containing a sample of mouse serum, and allowed to solidify. This solid basal layer was then layered with 1 ml of the above agar mixture containing 10^5 nucleated bone marrow cells. The cultures were incubated at 37°C in a humidified incubator. After 8 days, macrocolonies were counted[21].

To determine colony-stimulating factor (CSF) activity, individual mice were injected intraperitoneally with 10 μg *E. coli* K235 LPS. After 6 h, serum samples were prepared from mice. These serum samples were individually assayed for their colony-stimulating activity at a final concentration of 2% in the assay mixture.

Radioimmunoassay for Serum Amyloid A Protein (SAA)

Mice were immunized with 50 μg *E. coli* K235 LPS intraperitoneally and bled after 18 h from the orbital sinus. Serum aliquots (10 μl) were incubated with formic acid at 37°C for 24 h, diluted with distilled water, frozen, and lyophilized before being assayed in duplicate on plastic microtitration plates coated with affinity purified antibodies to mouse amyloid A protein[22].

RESULTS

Genetics of Defective LPS Responses

We have examined first, the genetic control of B lymphocyte responses to LPS. F_1 hybrid mice between C3H/HeJ mice and a number of LPS responder strains all show intermediate responses to LPS[20]. However, a variable that seems to influence kinetic patterns of mitogenic responsiveness is the density of cells in culture. At a cell density of 5.0 x 10^6 cells/ml, the kinetic patterns of LPS responsiveness differ in F_1 hybrid and responder parent spleen cultures[20]. Parent cultures show a sharp peak of mitogenic responsiveness on day 2, whereas F_1 hybrid cultures show a broad plateau of maximal responsiveness on day 2 and 3[20]. When the density of cells in a responder parent culture is reduced to 1.25 x 10^6 cells/ml, the kinetic pattern of mitogenic responsiveness is similar to that observed in F_1 cultures at a cell density of 5.0 x 10^6 cells/ml or 2.5 x 10^6 cells/ml.

Autoradiographic studies have indicated that the number of LPS responsive cells in F_1 cultures is approximately half the number of responsive cells in parent cultures (Table 1). These data correlate with the kinetic patterns in the two types of cultures and imply that the number of responsive cells in a culture is an important parameter in determining the kinetic pattern. Thus, it seems that kinetic responses to LPS most likely are primarily a result of the culture conditions and are not directly under genetic control[20].

In order to determine the number of genes involved in the defective LPS response in C3H/HeJ mice, we have performed a number of backcross linkage analyses. In backcross progeny from three responder F_1 strains of mice - (C3H/HeJxCWB/13)F_1, (C3H/HeJ x C57BL/6J)F_1 and (C3H/HeJxBALB/c)F_1 - each backcrossed to the non-responder C3H/HeJ parent, responder and nonresponder progeny were observed in a ratio of 1:1[14,20,23]. These results suggest that the lack of LPS responsiveness in C3H/HeJ mice is controlled by a single locus, and the alleles at this locus are codominantly expressed. The intermediate mitogenic responsiveness of F_1 cultures results from approximately half the number of LPS responsive cells in F_1 cultures as compared to the number of responsive cells in responder parent cultures. This finding suggests either a gene dosage phenomenon or allelic exclusion at the locus that controls LPS responsiveness[20].

TABLE 1

AUTORADIOGRAPHIC ANALYSIS OF SPLEEN CELLS TREATED WITH LPS[a]

Strain	Labeling time (hours)	% labeled cells
C3H/HeJ	24-32	<2
C57BL/6J	24-32	32
(C3H/HeJxC57BL/6J)F_1	24-32	16
C3H/HeJ	48-54	<2
C57BL/6J	48-54	16
(C3H/HeJxC57BL/6J)F_1	48-54	14

[a] These data are summarized from reference 20.

We have also compared the linkage relationships of the polyclonal and adjuvant responses to mitogenic responses in backcross (C3H/HeJxCWB/13)F_1 x C3H/HeJ mice[14,20,23,24]. Mitogenic, polyclonal, and adjuvant responsiveness to LPS all segregated together in the backcross progeny, demonstrating the expression of a single gene controlling each of these responses to LPS.

The inheritance of the defective mitogenic response to LPS in C57BL/10.Sc.Cr mice has been examined by Coutinho et al.[16,17]. F_1 hybrid mice between these nonresponder mice and responder C57BL/6J, BALB/c and C3H/Tif mice are all high responders.

In contrast to the defect in C3H/HeJ mice, the defect in C57BL/10.Sc.Cr mice appears to be inherited as an autosomal recessive trait[16,17]. In C3H/HeJ, the defect seems to be codominant[20,25]. F_1 hybrid mice between the two non-responder strains are also nonresponders to LPS. The lack of complementation suggests that these strains may possess defective alleles at the same locus.

In the progeny of backcrosses between C57BL/10.Sc.Cr and high responder mice, the LPS responsive trait segregates in a ratio of 1:1[16,17]. Thus the B cell defect in these mice also appears to result from a single gene defect.

Chromosomal Location of an LPS Response Gene

The use of a number of recombinant inbred (RI) strains of mice has led to the mapping of the LPS response gene in C3H/HeJ mice. Fourteen RI strains have been produced by inbreeding, beginning with randomly chosen pairs of mice from the F_2 of the cross between C3H/HeJ and C57BL/6J progenitor strains[18]. The data presented in Table 2, summarize the *in vitro* mitogenic response, and the *in vivo* adjuvant response to various concentrations of LPS.

TABLE 2

MITOGENIC AND ADJUVANT RESPONSES TO LPS IN BxH RECOMBINANT INBRED STRAINS[a]

	Mitogenic response (Cpm x 10^{-3})[b]			
Strain	No LPS	25 µg	Serum Titers[c]	LPS Response
C3H/HeJ	4.4	5.6	<10/10	Low
C57BL/6J	3.8	51.5	<10/1,280	High
BxH-2	1.1	5.1	<10/10	Low
BxH-3	1.1	3.1	<10/20	Low
BxH-4	3.7	8.3	10/20	Low
BxH-5	3.7	50.8	<10/1,280	High
BxH-6	3.3	4.5	<10/40	Low
BxH-7	2.2	3.0	10/80	Low
BxH-8	1.2	2.5	<10/10	Low
BxH-9	3.7	5.3	<10/10	Low
BxH-11	2.2	43.1	10/320	High
BxH-12	2.8	5.6	<10/10	Low
BxH-14	3.9	46.6	10/1,280	High
BxH-18	3.1	5.4	10/80	Low
BxH-19	3.5	44.5	10/320	High

[a] The data are summarized from reference 18.

[b] Mitogenic assays were performed as described in reference 18.

[c] Two figures are presented for the serum titers. The first represents the titer of the pre-immunization serum, and the second represents the titer 7 days after immunization. The serum titers are presented as a reciprocal of the end point dilution. The initial dilution was always 1:10.

BxH strains 2,3,4,6,7,8,9,12 and 18 all showed low mitogenic responses to LPS, as well as small adjuvant responses[18]. However, the BxH strains 5,11,12,14 and 19 all possess the LPS response characteristics of the high responder C57BL/6J parent. Not shown are data for the BxH-10 strain. These are high responders to LPS in both response assays[19].

In an initial survey of the expression of a number of genetic traits in the RI strains, we have observed that 13 of the 14 BxH strains exhibited concordant inheritance between the expression of the major urinary protein in mice and LPS responsiveness (Table 3).

Only a single strain, BxH-18, possesses a recombinant phenotype with respect to LPS responsiveness and _Mup-1_ (Table 3). This degree of concordance is formally significant ($p<0.01$), suggesting linkage of the two characters.

We have examined the expression of _Mup-1_ in 70 backcross (C3H/HeJxC57BL/6J)F_1 x C3H/HeJ mice. All progeny mice exhibit either the _Mup-1_^{a}/_Mup-1_^{a} homozygous or _Mup-1_^{a}/_Mup-1_^{b} heterozygous phenotypes. The phenotypes of individual mice were first determined by gel electrophoresis of urine samples. The mice were then sacrificed and the mitogenic response of B lymphocytes to LPS was determined in spleen cell cultures. Of the 70 backcross mice, 33 exhibited a _Mup-1_^{a}/_Mup-1_^{a} phenotype and did not support an _in vitro_ mitogenic response to LPS. Another 30 backcross mice exhibited a _Mup-1_^{a}/_Mup-1_^{b} phenotype and supported _in vitro_ mitogenic responses to LPS. Three _Mup-1_^{a}/_Mup-1_^{a} homozygotes supported the LPS response, whereas four _Mup-1_^{a}/_Mup-1_^{b} heterozygotes did not respond to LPS. The latter two cases are presumably recombinants. These data confirm genetic linkage between the expression of the

TABLE 3

SUMMARY OF THE MAJOR URINARY PROTEIN (MUP-1) CHARACTERISTICS OF RECOMBINANT INBRED C57BL/6J x C3H/HeJ LINES[a]

RI LINE	Mup-1	LPS (responder/ nonresponder)
BxH-2	a	NR
BxH-3	a	NR
BxH-4	a	NR
BxH-5	b	R
BxH-6	a	NR
BxH-7	a	NR
BxH-8	a	NR
BxH-9	a	NR
BxH-10	b	R
BxH-11	b	R
BxH-12	a	NR
BxH-14	b	R
BxH-18	b	NR
BxH-19	b	R

[a] Parental Types: C57BL/6J, Mup-1^{b}; C3H/HeJ, Mup-1^{a}.

Mup-1 locus and the locus responsible for LPS responsiveness, which we have termed Lps (Table 4).

It has been shown previously that Mup-1 is located on chromosome 4 of the mouse linked to the brown coat color (b) locus[26]. Mup-1 is located on the centromeric side of b. The locus for polysyndactyly (Ps) is also found on chromosome 4 on the distal side of b[27]. We have determined the location of Lps on chromosome 4

TABLE 4

SUMMARY OF BACKCROSS LINKAGE ANALYSIS[a]

Genotype	Number of high responder mice (Lps^n/Lps^d)	Number of low responder mice (Lps^d/Lps^d)
$Mup\text{-}1^a/Mup\text{-}1^a$	3	33
$Mup\text{-}1^a/Mup\text{-}1^b$	30	4

[a] (C3H/HeJxC57BL/6J)F_1 mice were backcrossed to C3H/HeJ and the progeny were tested for electrophoretic variants of the major urinary protein. C3H/HeJ is $Mup\text{-}1^a$ and C57BL/6J is $Mup\text{-}1^b$. Backcross mice were either homozygous ($Mup\text{-}1^a/Mup\text{-}1^a$) or heterozygous ($Mup\text{-}1^a/Mup\text{-}1^b$) at this locus. We use the locus symbol Lps, with the mutant allele of C3H/HeJ designated as Lps^d and the normal allele of other strains designated Lps^n. These data are from reference 19.

relative to Mup-1 and Ps in a three-point cross. F_1 hybrid mice from C3H/HeJxC57BL/6By-Ps parents were backcrossed to C3H/HeJ. The segregation of phenotypes for Mup-1, Lps, and Ps markers in the backcross progeny is shown in Table 5. The location of the Lps locus between the Mup-1 and Ps locus is consistent with the four recombinant phenotypes which were found[19].

The results of the three-point cross establish the gene order and approximate recombination frequencies as follows: Mup-1 - (0.06 ± 0.02) - Lps (0.13 ± 0.07) - Ps. Previous studies have mapped the Mup-1 and Ps genes on the centromeric and distal sides of brown (b), respectively[17,19]; recombination frequencies were 0.07 and 0.08 for the two intervals. Thus, Lps is probably near the b locus.

TABLE 5

SEGREGATION OF Mup-1, Lps, AND Ps MARKERS IN BACKCROSS C3H/HeJxC57BL/6By-PS)F_1xC3H/HeJ MICE[a]

Region of recombination	Genetic Locus Mup-1	Lps	Ps	No. of Mice
None	a	b	+	15
	b	n	Ps	10
Mup-1-Lps	a	d	Ps	2
	b	n	+	2
Lps-Ps	a	n	Ps	1
	b	d	+	1
Mup-1-Lps-Ps	a	n	+	0
	b	d	Ps	0
Total				31

[a] The genotypes for Mup-1, Lps, and Ps markers, respectively, are: C3H/HeJ, a d +/a d +; C57By/6By-Ps, b n Ps/b n +; F_1, a d +/b n Ps. Backcross mice were first typed for Mup-1 and Ps, and then spleen cultures were assayed for mitogenic responsiveness to LPS. These data are summarized from reference 19.

These genetic studies indicate that the low responsiveness of C3H/HeJ mice to LPS is determined by a gene that maps between Mup-1 and Ps on chromosome 4.

Recent studies with the C57BL/10.Sc.Cr strain also show that the defective LPS response gene is linked to the expression of Mup-1 (A. Coutinho and T. Meo, unpublished data). It is likely, therefore, that the defective genes in C3H/HeJ and C57Bl/10.Sc.Cr strains are found in the same genetic locus.

Expression of the LPS Locus in Nonlymphoid Cell Types

Our studies on the effect of LPS on B lymphoctyes have defined the general reactions to LPS in mice that constitute the endotoxic response. In addition to its resistance to the mitogenic, polyclonal, and adjuvant effects of LPS, C3H/HeJ mice also show resistance to a variety of non-immunologic effects such as toxicity[15], non-specific resistance to bacterial infections[11], macrophage activation[10], and increases in levels of various serum components such as colony stimulating factor (CSF)[12], and the acute phase serum amyloid protein (SAA)[21,22]. The major problem in attempting to correlate the genetic control of LPS responses in B lymphocytes to those involving the interaction of LPS with other cell types is the difficulty in performing more than one type of LPS response assay in individual animals. We have examined three non-immunologic responses induced by LPS, utilizing 12 of the recombinant inbred strains of mice derived from C3H/HeJ and C57BL/6J parental strains, and a backcross linkage analysis. The first response is a hypothermia induced by LPS when mice are placed in an environment below 25°C, which is in contrast to a hyperthermia when when the temperature is greater than 30°C[34]. The mechanisms regulating these fluctuations in body temperature are not known. The second response is the production of a colony stimulating factor (CSF) which stimulates granulocyte and macrophage colony formation by mouse bone marrow cells in agar cultures. LPS does not elevate serum CSF levels in C3H/HeJ mice as in other strains of mice[12]. The third LPS response examined is the production of a serum precursor SAA[28] from the secondary amyloid protein AA[22].

All recombinant inbred strains that express *Lps*[d] (BxH-2,3,4,6, 7,8,9,12) do not respond to an injection of LPS by hypothermia, or by increases in serum levels of either CSF or SAA (Table 6).

TABLE 6

EXPRESSION OF *Lps* AND THE ELEVATION OF SAA LEVELS BY *E. coli* K235 LPS IN RI STRAINS[a]

Strain	LPS	SAA (μg/ml)
BXH-2	d	<2
BXH-3	d	4
BXH-4	d	13
BXH-5	n	687
BXH-6	d	< 2
BXH-7	d	8
BXH-8	d	< 2
BXH-9	d	6
BXH-11	n	47
BXH-12	d	4
BXH-14	n	828
BXH-19	n	743
C3H/HeJ	d	< 2
C57BL/6J	n	401

[a] Individual mice from each strain were injected i.p. with 50 μg *E. coli* K235 LPS. After 18 h serum samples were prepared. The SAA concentrations were determined in a radioimmune assay (Materials and Methods). These data are summarized from reference 21.

The Lps^{n} strains (BxH-5,11,14, and 19) all show hypothermal and CSF responses to LPS. In three of these RI strains, BxH-5,14, and 19, LPS induced large increases in SAA levels. However, one Lps^{n} strain, BxH-11, shows only a slight SAA response to LPS, which may reflect the involvement of a genetic locus in the synthesis of SAA that is distinct from LPS. This locus is apparently altered in BxH-11 mice.

A backcross linkage analysis has been used to examine the involvement of Lps in those same three LPS-induced responses. F_1 (C3H/HeJxC57BL/6J) mice were backcrossed to C3H/HeJ mice and the individual progeny were examined in a number of ways. First, backcross progeny with a homozygous Mup-1^{a}/Mup-1^{a} phenotype did not show either hypothermal responses to an injection of LPS or increases in serum CSF or SAA. However, backcross progeny with heterozygous Mup-1^{a}/ Mup-1^{b} phenotypes were LPS responders in each of the three assays. Several recombinant phenotypes were observed, as would be expected upon our estimation of the recombination frequency between Mup-1 and Lps. Thus, the expression of Mup-1 seems to be an accurate genetic marker for the linked locus, Lps. To verify this, we have kept mice for 3 weeks after a single injection of LPS and subsequent analysis of CSF or SAA levels, and then assayed mitogenic responses to LPS in spleen cultures. We know the mitogenic response to be a direct assay of the expression of the Lps locus. With the exception of several presumed recombinant backcross progeny, all Mup-1^{a}/Mup-1^{a} mice were Lps^{d}, whereas Mup-1^{a}/ Mup-1^{b} mice were Lps^{n}. Therefore, there was concordance in the expression of Mup-1 and Lps in each LPS response assay[21].

DISCUSSION

Our genetic studies have shown that the expression of the Lps locus is required for the initiation of LPS responses in many different cell types. To initiate a specific biochemical response, LPS must interact with a cellular structure. The simplest interpretation of these results is that the Lps locus is involved in the production of an LPS receptor or in the activation of an enzyme through which the LPS receptor exerts its effect.

In an attempt to find an LPS receptor, Coutinho *et al.*[16,17] have prepared antisera by immunizing rabbits with LPS-high responder (C3H/Tif) spleen cells, and then absorbed the antiserum in LPS-nonresponder (C3H/HeJ) mice. The resulting antisera detect surface markers on lymphoid cells from all LPS-responsive mouse strains tested, and fail to detect these markers on the two LPS-nonresponder strains, C3H/HeJ and C57BL/10.Sc.Cr[16,17]. The binding of this antisera to cells from high responder mice can be inhibited by lipid A, and the antisera itself appears to be mitogenic for B lymphocytes from LPS-responder strains[16,17]. Such antisera have the properties that might be expected if they have specificity for a surface receptor for lipid A. If the Lps locus codes for such a receptor structure, we would expect to find this receptor on many cell types. Such analyses have yet to be performed.

There are several interesting features of the genetic system that appear to control the response of mice to lipid A. First, lipid A directly activates B lymphocytes and is, thus, providing the cell with the signals normally given by interaction with specific antigen and helper T cells. Lipid A is a tool which can be used to dissect the biochemical control of the mechanism controlling the inducion of antibody synthesis. Second, the interaction of lipid A with other cell types in mammals leads to endotoxic reactions that can be extremely harmful. Humoral antibody to the O-antigens of the LPS molecule has little effect in protecting against the endotoxic effects of the lipid A moiety. Knowledge of the Lps locus may lead to the development of inhibitory agents that block either the interaction of lipid A with cells, or the initiation of the subsequent biochemical response, thereby preventing the onset of endotoxic reactions.

Because of their very close association with host organisms, bacteria provide a constant source of exogenous substances that can interact with host tissue. The finding that many bacterial wall constituents act as effector agents for the immune system of mammals may be of general importance for the modulation of immune responsiveness to infection and disease. Immunologic activities resulting from the nonspecific activation of lymphocytes and monocytes by bacterial constituents may have protective or destructive effects on host tissue. The protective effects are revealed experimentally by an increased resistance to subsequent infection by bacteria and fungi, and also the growth of transplanted tumors[3]. The destructive effects of chronic bacterial infections may be seen in inflammatory responses such as actinomycosis, perodontitis and nocardiosis; in these cases there are sites of intense lymphocytic and monocytic activity and host tissue destruction. The mechanism of initiation of these protective and destructive processes has many implications in health and disease. Molecules, such as LPS, released from the cell wall of invading pathogens, may interact directly with lymphocytes and monocytes to induce effector cell function. It is the expression of these immune functions that leads to the phenomenon of nonspecific host resistance to infection, or to the appearance of chronic inflammation. Lymphocytes and monocytes may possess surface receptors for a variety of bacterial cell wall constituents. These genetic studies may serve as a model, not only for determining how the host responds to molecules such as lipid A, but for investigating how other molecules such as the lipoprotein and peptidoglycan structures from cell walls, also influence the immune system.

ACKNOWLEDGEMENTS

This work is a result of a collaborative effort with many workers in the past four years. In particular I gratefully acknowledge the contributions of Drs. R. Riblet, B.A. Taylor, M. Largen, K. McAdam and Kathleen Kelly in furthering this work. The experiments have been supported by Public Health Service Grant AI 13383 from the National Institute of Allergy and Infectious Diseases, and Grant 1-469 from the National Foundation. James Watson is supported by a Research Career Developmental Award AI-00182 from the National Institute of Allergy and Infectious Diseases.

REFERENCES

1. Watson, J., Kelly, K. and Largen, M. (1978) in Genetics of Infection and Immunity, Friedman H. and Prier J. eds. Temple University Press, Philadelphia, pp. 25-38.

2. Braun, V. and Bosch, V. (1972) Eur. J. Biochem. 28, 51-69.

3. Chedid, L., Audibert, F. and Johnson, A.G. (1978) Prog. All. 5, 1-40.

4. Landy, M. and Baker, P.J. (1966) J. Immunol. 97, 670-679.

5. Rudbach, J.A. (1971) J. Immunol. 106, 993-1001.

6. Skidmore, B., Chiller, J., Morrison, D. and Weigle, W. (1975) J. Immunol. 114, 770-776.

7. Andersson, J., Möller, G. and Sjöberg, O. (1972) Cell. Immunol. 4, 381-393.

8. Von Eschen, K.B. and Rudbach, J.A. (1974) J. Exp. Med. 140, 1604-1614.

9. Sultzer, B.M. (1968) Nature, 219, 1253-1254.

10. Ruco, L.P. and Meltzer, M.S. (1978) J. Immunol. 120, 329-334.

11. Chedid, L.M., Parant, M., Damais C., Parant F., Juy, D. and Galelli, A. (1976) Infect. Immun. 13, 722-727.

12. Apte, R.N. and Pluznik, D.H. (1976) J. Cell. Physiol. 89, 313-323.

13. Westphal, O. (1977) in Microbiology (1977) Schlessinger, D. ed. American Society for Microbiology, Washington D.C. pp. 220-230.

14. Watson, J. and Riblet, R. (1974) J. Exp. Med. 140, 1147-1161.

15. Sultzer, B.M. (1976) Infect. Immun. 13, 1579-1584.

16. Forni, L. and Coutinho, A. (1978) Eur. J. Immunol. 8, 56-62.

17. Coutinho, A., Forni, L. and Watanabe, T. (1978) Eur. J. Immunol. 8, 63-66.

18. Watson, J., Riblet, R. and Taylor, B.A. (1977) J. Immunol. 118, 1604-1610.

19. Watson, J., Kelly, K., Largen, M. and Taylor, B.A. (1978) J. Immunol. 120, 422-424.

20. Kelly, K. and Watson, J. (1977) Immunogenetics, 5 75-84.

21. Watson, J., Largen, M., McAdam, K.P.W.J. and Taylor, B.A. (1978) J. Exp. Med. 147, 39-49.

22. McAdam, K.W.P.J. and Sipe, J.D. (1976) J. Exp. Med. 144, 1121-1127.

23. Watson, J. and Riblet, R. (1976) J. Immunol. 114, 1462-1468.

24. Skidmore, B.J., Chiller, J.M. Weigle W.O., Riblet, R. and Watson, J. (1976) J. Exp. Med. 143, 143-150.

25. Glode, L.M. and Rosenstreich, D.L. (1976) J. Immunol. 117, 2061-2067.

26. Hudson, D.M., Finlayson, J.S. and Potter, M. (1967) Gen. Res. 10, 195-198.

27. Johnson, D.R. (1969) J. Embryol. Exp. Morphol. 21, 285-296.

28. Rosenthal, J., Franklin, E.C., Frangeine, B. and Greenspan, J. (1976) J. Immunol. 116, 1415-1418.

Rudbach/Baker, eds. Immunology of Bacterial Polysaccharides

IDIOTYPES OF RABBIT ANTISTREPTOCOCCAL ANTIBODIES: PROBES FOR INHERITANCE AND IMMUNE REGULATION

MARTIN L. YARMUSH AND THOMAS J. KINDT
Laboratory of Immunogenetics, Building 8, Room 100,
National Institute of Allergy and Infectious Diseases,
Bethesda, Maryland 20014

ABSTRACT

Upon immunization with streptococcal vaccines, certain rabbits produce large amounts of antibodies which may exhibit molecular restriction. These antibodies are directed against cell-surface antigens of the streptococcus, usually components of the group-specific carbohydrates. Idiotypes of homogeneous rabbit antibodies directed against the streptococcal polysaccharide antigens have proven useful markers in studies of antibody V region inheritance. Anti-idiotype antibodies directed against heavy chain specificities have been used to demonstrate inheritance of antibody V_H regions as well as linkage to the V_H allotypes. Recent studies demonstrating that injection of anti-idiotype antibodies can induce formation of cognate specificities suggest that idiotype markers may perform a mediatory role in the regulation of antibody synthesis.

INTRODUCTION

The availability of myeloma proteins in man and mouse has greatly facilitated the elucidation of immunoglobulin structure and has provided much insight into the genetic mechanisms controlling various aspects of the immune response. Although rabbit myeloma proteins have not been found, an appropriate substitute was made available with the observation of Krause and co-workers that certain rabbits immunized with streptococcal vaccines produced large amounts of homogeneous antibodies[1]. The methodology required for the isolation of large amounts of these homogeneous components was soon developed[2], and extensive serologic and structural studies of rabbit antibodies became a possibility.

The major importance of the rabbit in immunogenetic studies is based upon the variety of allotypic markers and their strategic locations on the different segments of the immunoglobulin molecule[3]. Genetic studies on idiotypes in the rabbit have been aided by this well described allotypic network and a much valued contribution of the streptococcal immunization system has been in the characterization of idiotype inheritance in the rabbit[4-10].

This presentation is divided into four areas. The first and second provide background information on the chemistry and immunochemistry of the relevant polysaccharide antigens of streptococci, immunization procedures, and isolation of homogeneous antibodies; the third deals with the genetics of rabbit idiotypes and the fourth concerns the possible implication of idiotypes as mediators of immune response regulation.

CHEMISTRY OF STREPTOCOCCAL POLYSACCHARIDE ANTIGENS

Recent chemical studies have revealed structural details and relationships among the group-specific polysaccharide antigens of group A, A-variant, and C streptococci[11-13]. The proposed structures for these group-specific polysaccharides are shown in Fig. 1. Group A-variant carbohydrate consists of a linear homopolymer of

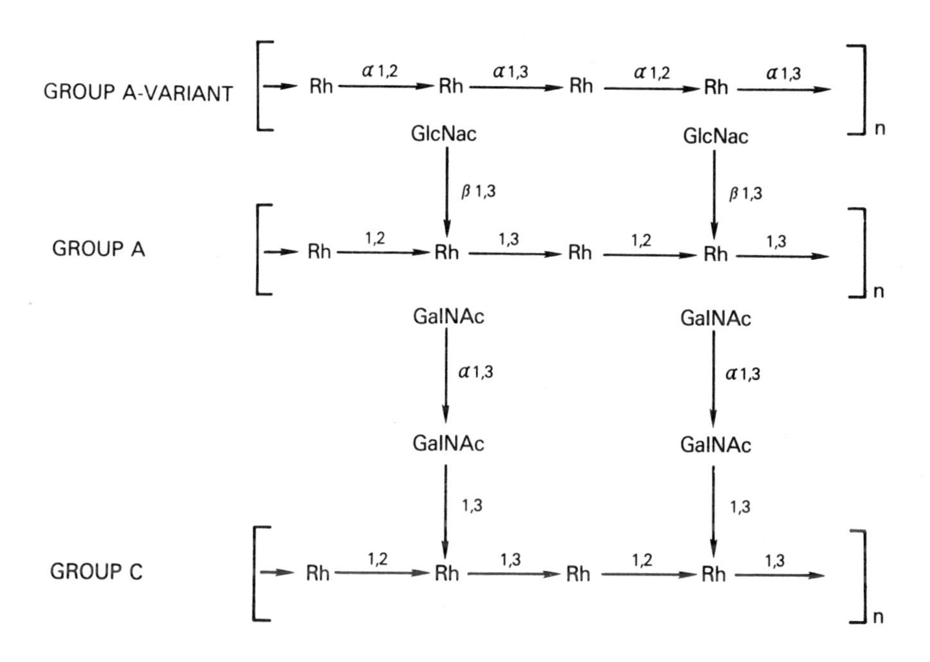

Fig. 1. Structures of the repeating oligosaccharide units for the group-specific carbohydrates from streptococci A-variant, A and C.

alternating α-1,2- and α-1,3-linked rhamnose residues. Group A and C polysaccharides contain, in addition to the rhamnose homopolymer, amino sugar side chains: β-linked N-acetylglucosaminosyl and, α-linked di-N-acetylgalactosaminosyl residues in group A and group C, respectively. Immunochemical studies have indicated that the amino sugar side chains are the immunodominant determinants of the group A and C polysaccharides. For example, complete inhibition of binding of ^{125}I streptococcal group C polysaccharide to various antistreptococcal group C sera was observed with a hapten, 3-0-α-GalNAc-GalNAc, isolated from a limited acid hydrolysate of the group C polysaccharide (Table 1)[12]. Similarly, binding of antibodies to group A polysaccharide can be inhibited by monomeric GluNAc and more effectively by the disaccharide 3-0-α-N-acetylglucosaminosyl-rhamnose. Further details concerning the immuno-chemistry and chemistry of the streptococcal group-specific carbohydrates can be found in reviews by Krause (1963)[14], McCarty (1971)[15] and Coligan et al., (1978)[16].

TABLE 1

INHIBITION OF BINDING OF GROUP C CARBOHYDRATE TO ANTI-GROUP C CARBOHYDRATE SERA BY GalNAc - GalNAc

Rabbit Donor of Antisera[a]	Inhibition by	
	GalNAc - GalNAc (134 nmole/assay tube)	GalNAc - GalNAc (67 nmole/assay tube)
	(%)	(%)
4811	99	98
4838	98	97
4844	98	92
4588	95	92
4827	93	76
4664	91	82
4813	85	78
4801	82	63
4798	38	25

[a] The antisera were obtained from rabbits during various stages of the immunization procedure described by Krause[17].

IMMUNIZATION PROCEDURES AND ISOLATION OF HOMOGENEOUS ANTIBODIES

Microbial polysaccharides are excellent antigens and will frequently produce high concentrations of homogeneous antibodies if administered as constituents of the bacterial cell. The group-specific streptococcal polysaccharides are structural elements of the bacterial cell wall in contrast to what is found in pneumococci where the analogous polysaccharide antigens reside in the capsule surrounding the cell wall.

Immunization is performed intravenously with a heatkilled pepsin-digested streptococcal suspension, the rhamnose content of which is 0.3-0.4 mg/ml[17]. A typical immunization schedule is illustrated in Table 2. Few rabbits respond with a major homogeneous component during the first four weeks of immunization. If, however, the rabbits are rested for several months and then given a second series of immunizations, a substantial increase in antibody production and a more frequent occurrence of homogeneous antibodies is observed. In order to obtain sufficient antibody for structural study, a multiple exchange transfusion protocol was developed[2]. By application of this method, up to 20 grams of specific antibody have been obtained from individual rabbits.

Antistreptococcal sera which contain a single major homogeneous component also contain, in most cases, minor populations of other antibodies. Therefore, it is necessary to fractionate antistreptococcal sera in order to obtain pure homogeneous components. A variety of techniques have been used to purify individual antibodies (affinity chromatography, electrophoretic and electrofocusing techniques, ion exchange chromatography, etc.) and occasionally a combination of these methods is required.

TABLE 2

IMMUNIZATION SCHEDULE FOR STREPTOCOCCAL VACCINES

Primary Series

	MON	TUES	WED	THURS
Week one	(bleed)	0.5 cc	0.5 cc	0.5 cc
Week two		1 cc	1 cc	1 cc
Week three		1 cc	1 cc	1 cc
Week four	(bleed)	1 cc	1 cc	1 cc
Week five	(bleed)		(bleed)	

3 - 5 Month Rest Period

Secondary Series

	MON	TUES	WED	THURS
Week one	(bleed)	0.25 cc	0.25 cc	0.25 cc
Week two		1 cc	1 cc	1 cc
Week three	(bleed)	1 cc	1 cc	1 cc
Week four	(bleed)		(bleed)	

The use of fractionation procedures is typified by recent work from our laboratory in which 10 different antibodies were isolated from the serum of a single rabbit by immunoadsorbent chromatography followed by preparative agarose block electrophoresis[18] (Fig. 2). Homogeneity of the light chains of each fraction was assessed by alkaline urea polyacrylamide gel electrophoresis. Six were found to exhibit a single light chain band.

Although during the initial steps of purification, electrophoresis patterns on cellulose acetate membranes are useful as

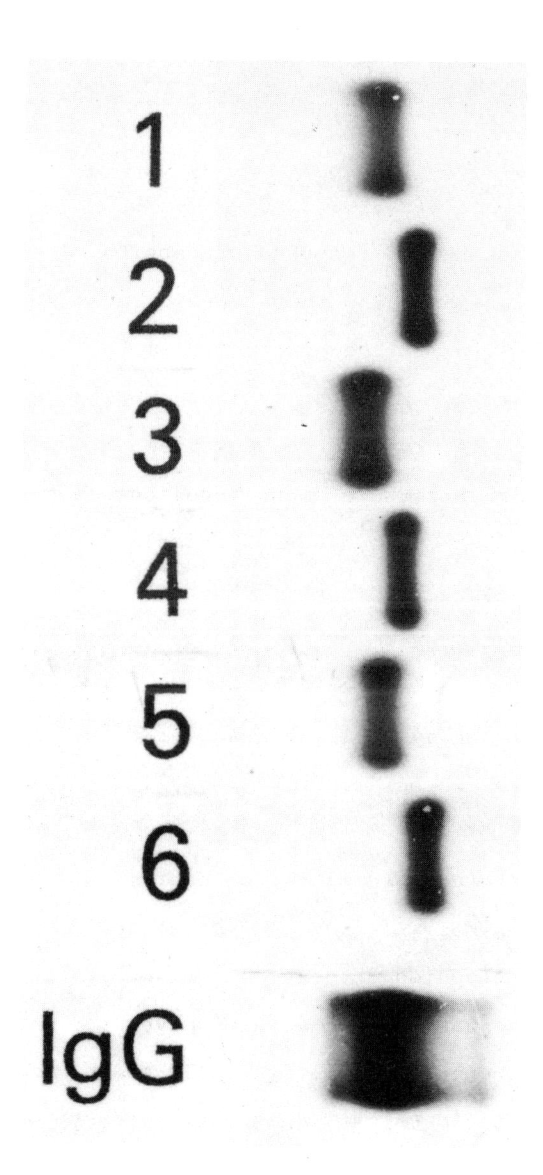

Fig. 2. Electrophoresis patterns on cellulose acetate of 6 of the 10 antibodies isolated from the immune plasma of rabbit 4295 compared to the total IgG pattern.

guides for uniformity, the two most rigorous criteria for antibody homogeneity are a single band in isoelectric focusing analysis and a single variable region amino acid sequence.

GENETICS OF THE ANTISTREPTOCOCCAL IDIOTYPES

Early studies on the response of rabbits to streptococcal hyperimmunization indicated that genetic factors influence the magnitude as well as the heterogeneity of the response[19,20]. By selective inbreeding, Eichmann *et al.*[19] were able to demonstrate i) the existence of high and low responder traits and ii) that the progeny of restricted or monoclonal responders contained a significantly higher percentage of animals producing a restricted or monoclonal response than that observed in progeny of random rabbits. Extending these observations, Eichmann and Kindt reported that the expression of individual antistreptococcal antibodies showed a strong familial association and, thus, demonstrated that the expression of antibody idiotypes is under genetic control[4].

Two aspects of the antistreptococcal idiotypic system should be mentioned. First, in the study cited above, heterologous idiotypic antisera prepared in guinea pigs were used to detect idiotypes. Subsequent idiotypic studies have used homologous anti-idiotype sera prepared in allotypically matched rabbits. Antisera of the latter type detect a more restricted set of specificities than do heterologous reagents. Second, although several other studies of idiotypic inheritance have employed antisera raised against whole antibody fractions[24], studies to be described here involve anti-idiotype sera raised against single homogeneous antibodies. The specificity of the latter reagents allows for the distinction between idiotypic identity and partial crossreactivity, and also permits identification of structural correlates of idiotypic specificities.

Several studies have confirmed and extended the initial observations on the inheritance of idiotypy in the rabbit[5-10]. For example, in one study 58% of the related rabbits expressed the proband idiotype while it was detected in only 1% of the streptococcal antisera from nonrelated rabbits[5,7]. This particular study used a test system involving inhibition of the binding reaction between the idiotypic antiserum and a crossreactive streptococcal antibody. It was found that the idiotype was preferentially expressed in rabbits which contained the same group a allotype as that of the proband antibody. Greater than 60% of the rabbits containing the a2 allotype expressed the idiotype, whereas only 4% of the related rabbits lacking this allotype exhibited idiotypic reactivity. In another study, structural examination of idiotypically crossreactive antibody molecules indicated that the L chains of the crossreactive molecules had the same N-terminal sequence as the proband antibody[21].

While these studies show the value of the idiotype as a genetic marker several unsuccessful attempts to demonstrate idiotypic inheritance have also been reported. In one of the studies cited above, when the frequency of expression of a second idiotype was studied, only 2 of more than 100 related rabbits produced antibodies with idiotypic crossreactivity[5]. Similarly, in another laboratory, idiotypic crossreactivity among related and unrelated rabbits was not observed for several homogeneous antipneumococcal antibodies[22]. Oudin and Bordenave also found low frequencies of idiotypic expression in their studies of anti-salmonella idiotypes and further observed that related rabbits had only a slightly higher frequency of crossreactivity[23,24].

Difficulties in detecting idiotypic inheritance may be attributable to a variety of factors. For example, the choice of assay method and reagents used may strongly influence the results obtained[25]. Moreover, there is no assurance that a particular idiotype will be elicited in a given immune response even when the appropriate genes are present in the animal genome; and if expressed, it may endure for only a brief period of time[26-29]. But perhaps more important is the fact that the demonstration of idiotypic expression involves the detection of a complex set of determinants[25].

A possible procedure for dissecting the idiotypic entity (which may in fact have physiological significance) involves separation of the antigen-binding site determinants from other variable region non-binding site determinants[30,31]. Recently, in an attempt to analyze idiotypic determinants in this fashion, antibodies from idiotypic antisera directed to homogeneous anti-streptococcal antibodies were separated into two distinct fractions which were subsequently shown to have differential susceptibilities to hapten inhibition of idiotypic binding[32]. Fractionation was performed by passing an idiotypic antiserum through an immunoadsorbent containing the antibody to which it is directed and then eluting the bound antibodies sequentially with GalNAc-GalNAc and 3M NH_4SCN in order to obtain two fractions. Although there appear to be antibodies in the latter fraction that are inhibited by high concentrations of hapten, the difference between the two fractions is apparent and indicates that such manipulations may yield two mutually exclusive anti-idiotypic antibody populations.

Another procedure for dissociating the idiotypic complex involves development of reagents specific for H and L chain determinants. Because most idiotypes require a specific H-L chain combination for their expression, genes at two unlinked loci must be coordinately expressed in order for an idiotype to be detected [23,33-35]. Thus, if a given idiotype could be serologically dissected into its respective H and L chain components the detection of the product of one variable region gene would no longer require the concommittant expression of the other.

In a recent study, an idiotypic reagent specific for determinants of a single H chain was prepared as outlined in Fig. 3[9]. H chains were isolated from a homogeneous antibody, 4135, and paired with an L chain pool from the total antistreptococcal fraction of a second rabbit. The H-L mixture was then injected into the rabbit from which the L chain pool was obtained in the expectation that the L chains would be nonimmunogenic. The antiserum raised was indeed specific for the H chain of homogeneous antibody 4135. The antiserum, designated 4135 HId (heavy chain idiotype), bound 4135 antibody and recombinant molecules containing 4135 H chains, but reacted neither with 4135 L chains or 4135 L chain recombinant molecules. When sera from related and unrelated rabbits immunized with group C streptococcal vaccine were tested for the presence of 4135 HId, 43% of sera from related rabbits with all possible combinations of group a allotypes were typed positive, whereas only 2 of 24 sera from unrelated rabbits showed

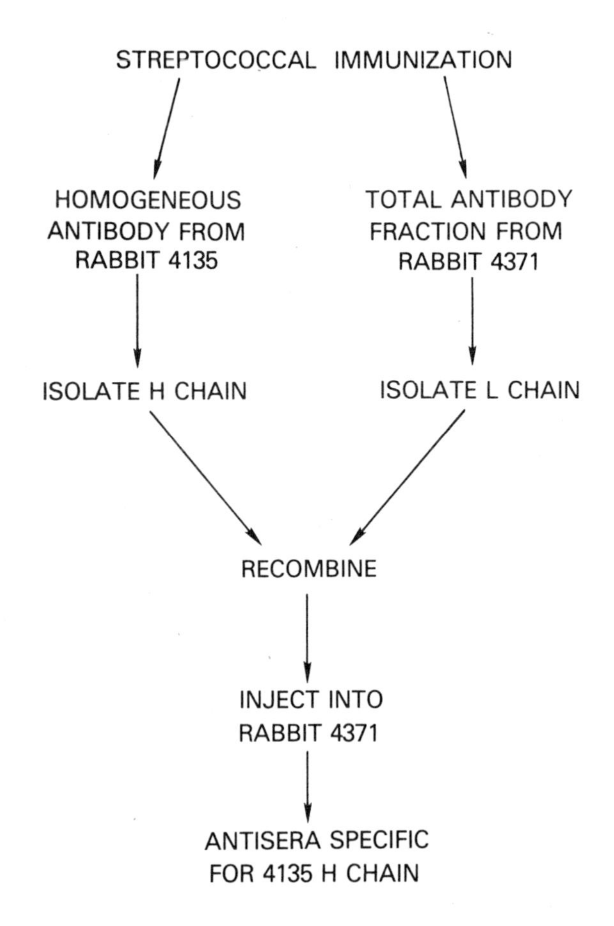

Fig. 3. Procedure for preparing an anti-idiotype serum specific for the H chain of a homogeneous antibody. Hybrid molecules containing the H chain of interest and L chains from an allotypically matched donor were injected into the L chain donor.

idiotypic reactivity.

In most instances, the idiotypic binding reaction between an antiserum and the antibody to which it was raised cannot be inhibited even by antibodies raised against the same antigen in

TABLE 3

DETECTION OF IDIOTYPICALLY CROSSREACTIVE ANTIBODIES BY INHIBITION OF HOMOLOGOUS BINDING

Idiotype[a]	Number of Sera Tested[b]	Number Positive	% Positive
3412	50	0	0
3521	60	2	3
4135	70	0	0
4153	55	0	0
4539	52	0	0
4135H	70	30	43
4539L	52	29	56

[a] All idiotypic antisera with the exception of 4135H and 4539L were shown to react with determinants expressed on intact antibodies and not on isolated H and L chains.

[b] All sera were from related rabbits which had been immunized with group C streptococcal vaccine.

related rabbits. Data shown in Table 3 indicate that crossreactions are not observed when homologous idiotypic systems are used to detect complex idiotypes. In these cases, idiotypic crossreaction was demonstrated by either binding reactions or by the use of heterologous, cross-idiotypic systems. However, in the case of 4135 HId and a light chain idiotype, 4539 LId[8], crossreactivity was demonstrated by the ability of sera to inhibit the homologous binding reactions (Table 3). Furthermore, examination of the allogroups[36] of the related rabbits expressing 4135 HId clearly indicates that aproximately 75% of the rabbits with the a3 allotype linked to the J haplotype (a^3 x^{32} y^- n^- f^{72} g^{74} d^{11} e^{15}) express the idiotype (Fig. 4). Thus, by restricting the specificity of an idiotypic antiserum to a single chain, the detection of an idiotype has been simplified so that homologous inhibition can be used. Furthermore, when the related rabbits tested, expressed the same allogroup as the proband antibody, the majority was found to express the idiotype.

It was of considerable importance to determine the molecular associations between the HId and the group a and group d allotypes, because the majority of rabbits typed positive for HId were heterozygous with respect to their a and d allotypes. By isolating the HId positive molecules using an anti-4135 HId immunoadsorbent and serologically characterizing the isolated components, the molecular association of the idiotype with the a and d allotypes of the J allogroup (a3/d11) was established in several rabbits.

Certain precautions must be taken in the isolation of small amounts of antibody for serologic analysis. For easy reference, the isolation procedure is illustrated in Fig. 5. Briefly, IgG or an antibody fraction is radiolabeled and passed through an immunoadsorbent containing anti-idiotype antibodies. The column is washed extensively and the bound fraction eluted with 3M NH_4SCN in PBS. The eluted antibody is either dialyzed against PBS or applied to a desalting column equilibrated with PBS and then passed through a second immunoadsorbent containing an IgG pool. The second immunoadsorbent serves to remove rheumatoid factors, proteins which were non-specifically bound to the first column and antibody denatured by the elution procedure. Two points are worthy of mention. First, the initial immunoadsorbent should be prepared with an isolated antibody fraction. Whole antisera or

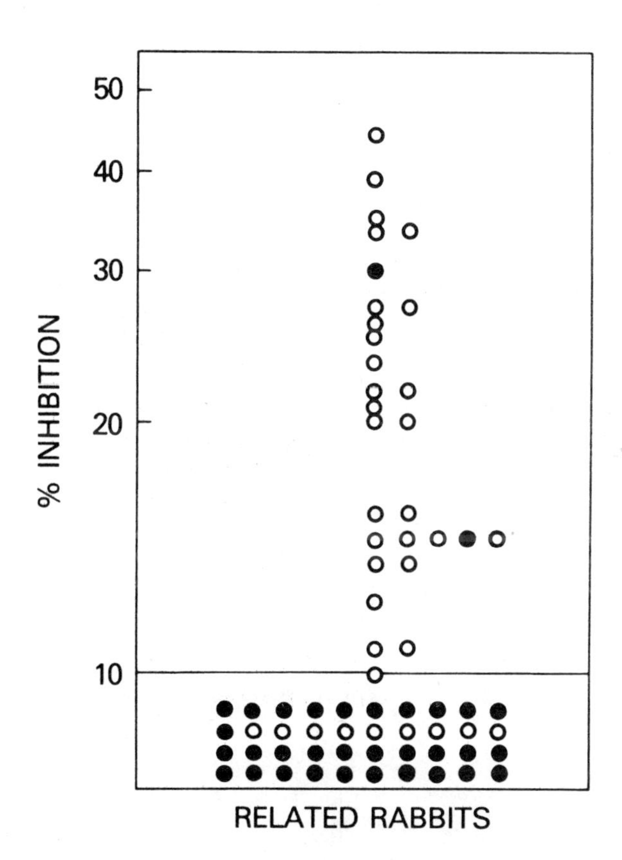

Fig. 4. Detection of 4135 HId in antistreptococcal sera from related rabbits expressing (o) or lacking (●) the allogroup of the proband rabbit (J). HId cross-reactions were detected by inhibition of the binding reaction between radiolabeled 4135 antibody and immobilized anti-HId. Inhibition values greater than 10% are considered positive.

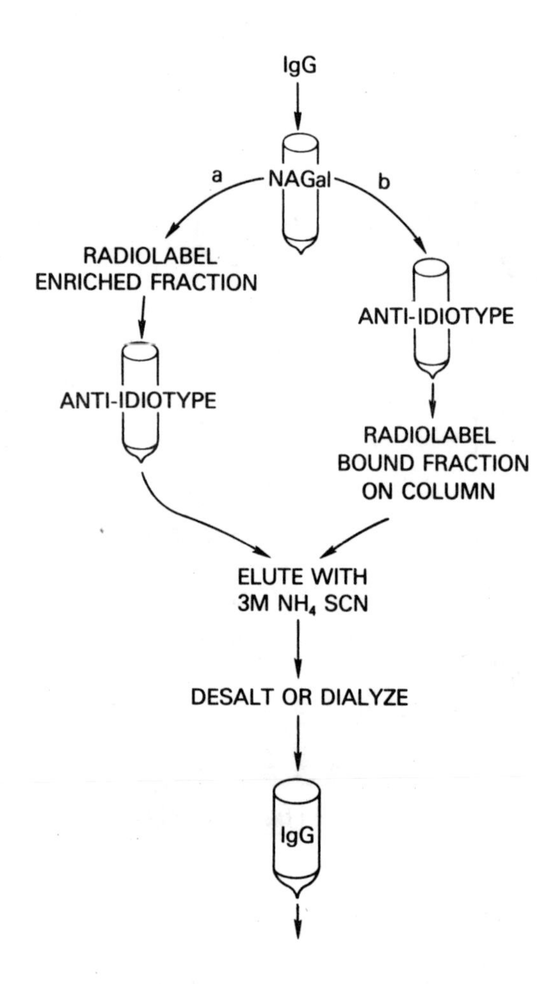

Fig. 5. Isolation of idiotypically crossreactive antibodies by immunoadsorbent chromatography. Designations a and b represent the alternative procedures described in the text.

IgG fractions usually contain antibodies with such high affinity for antigen that even if the column is subjected to extreme denaturing conditions, insufficient quantities of antibody are eluted. Second, because the final product is only a small percentage of the total IgG, it is advantageous to first fractionate the sera under question in order to enrich for the desired component before radiolabeling. Recently, we have had success with a technique that involves radiolabeling of antibodies bound to immunoadsorbent columns[37]. In this procedure, the total antibody fraction may be lightly radiolabeled and then the bound portion re-iodinated on the column without much loss of serologic reactivity.

Although idiotypes have been shown to be preferentially associated with antibodies with the same V region parameters as the proband antibody[5,7,9,21], several examples of identical or cross-reactive idiotypes on molecules with differences in other V region markers have been uncovered[6,38-41]. The first observation was made with a pair of antibodies isolated from rabbit 3521[38]. The antibodies had very similar affinities for antigen and similar electrophoretic mobilities while their light chains had identical N-terminal sequences and indistinguishable electrofocusing patterns. Surprisingly, however, one of the antibodies was found to lack group a allotypic determinants and to display determinants of the x or y locus, whereas the other antibody was characterized as having the a3 allotype. In spite of the difference in the V_H allotypes, the molecules were shown to have identical idiotypes by reciprocal inhibition of homologous binding analyses.

Since this initial observation, several similar examples have been reported. In another instance, idiotypic cross-reactivity was observed with antibodies 4153-I and 4153-II also isolated from a single rabbit[40]. Although both antibodies contained the a3 allotype and showed identical L chain N-terminal sequences, they behaved quite differently electrophoretically. Inhibition of binding studies using recombinant molecules showed that the light chains were idiotypically identical. Recent structural data have indicated that both of these L chains exhibit the same sequence for the first two hypervariable regions, yet differ with regard to two framework residues at positions 63 and 70 (Fig. 6)[42].

These observations of idiotypic identity among V_H regions differing in framework structure have assumed the focal point in challenges of models that depict V region genes as uninterrupted DNA sequences encoding hypervariable regions and framework residues, and they provide support for gene insertion theories and theories which call for directed somatic mutation[43-45].

IDIOTYPES AND REGULATION OF THE IMMUNE RESPONSE

Recently, two laboratories have demonstrated that injection of anti-idiotypic antibody raised to a total antibody fraction, followed by immunization with antigen, induced, in random rabbits, the expression of antibodies which share idiotypic specificities with the original set of idiotypes[46,47]. Intrigued by these results, we immunized three allotypically matched rabbits with purified anti-idiotype antibodies raised against homogeneous antibody 4136. Following a short rest period, the animals were immunized with streptococcal C vaccine and the sera were tested for inhibition of the homologous binding reaction between anti-4136 and 4136 antibody. Some preliminary results obtained using

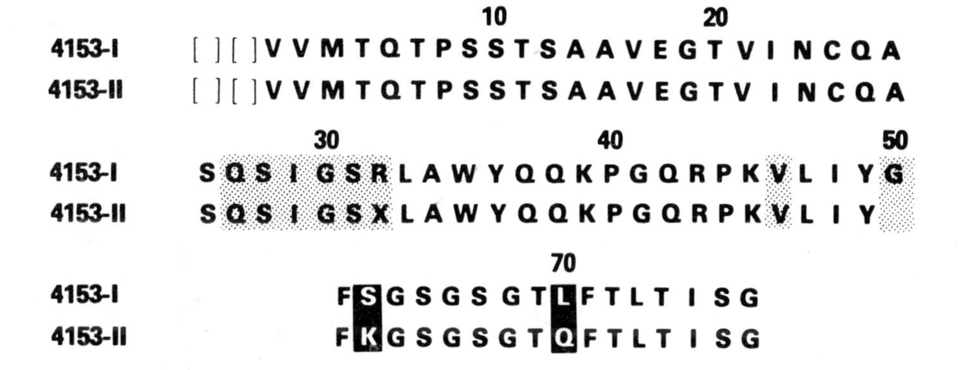

Fig. 6. Sequence data obtained for 4153 (b9) L chains. The numbering system is used to generate maximal homology to rabbit b4 L chains. Hypervariable regions are shaded and differences in framework sequence between the two antibodies are enclosed in black.

the sera from one of the three rabbits, 5374, are shown in Fig. 7. As can be seen, after four weeks of streptococcal immunization, components crossreactive with 4136 antibody appeared. These results suggest that by suitable manipulation (injection of anti-idiotype) one may induce an immune response against autologous suppressor components (cognate auto-anti-idiotypes) and thus allow expression of the desired idiotype. Amino acid sequence studies of 4136 antibody are currently in progress and hopefully will serve to elucidate the structural basis for the idiotypic crossreactivity.

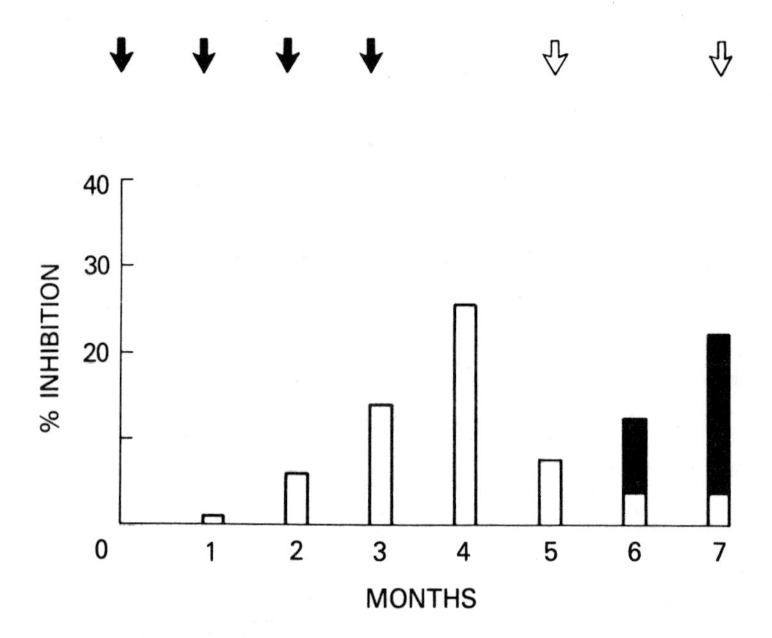

Fig. 7. Detection of idiotypically reactive molecules in the serum of rabbit 5374. Monthly injections of anti-idiotypic antibodies directed against 4136 antibody (black arrows) were followed by streptococcal immunization (white arrows). The white bars represent percent inhibition of the homologous binding reaction between 4136 antibody and anti-4136 by serum which was preadsorbed with streptococcal vaccine. The black bars represent percent inhibition of unadsorbed serum minus the inhibition obtained with adsorbed serum. It is likely that white bars represent the level of anti-(anti-4136) whereas the black bars represent anti-streptococcal antibodies idiotypically cross-reactive with 4136 antibody.

The recently advanced network theory describes the immune system as a self-regulatory unit[48]. When antigen is introduced, certain idiotypic communities expand beyond their equilibrium concentrations and thus complementary anti-idiotypes respond in order to relieve the stress. This concept of a web of idiotypes and anti-idiotypes acting to regulate the immune system has found support in recent experimental work. Auto-anti-idiotypic antibodies have been reported following immunization with the idiotype[49] or as spontaneously occurring products of immunization with the other antigens, with an allograft, or in the course of malignancy[50,51]. It has also been shown that the cellular compartment which recognizes idiotypic determinants contains both B and T cells and that the recognition phenomena displayed by these two cell populations may be different[50,52-57]. Furthermore, there is considerable experimental evidence for the inductive and suppressive capabilities of different anti-idiotypic components[58,59]. Indirect evidence for the control of antibody synthesis by anti-idiotypes is given by reports that individual rabbits during the course of immunization produce antibodies with similar or identical idiotypes. Idiotypically-crossreactive antibodies with differences in antigen binding specificity[24,60] and differences in V region allotypes have been observed[38,39,61].

However, sufficient direct evidence of a functional network has yet to be obtained and many unanswered questions remain: Why are natural auto-anti-idiotypic responses, except in rare cases, difficult or impossible to detect? What determines the suppressive _vs_. stimulatory activities of auto-anti-idiotypes (immunoglobulin class)? Is there binding site _vs_. non-binding site recognition?

CONCLUSION

In summary, the use of streptococcal polysaccharides as antigens has provided an excellent means for analyzing various aspects of the immune response. The success of this analysis has been due to the well characterized structural and serological properties of the streptococcal carbohydrate moieties and of the antibodies elicited by streptococcal immunization. Further definition of the number and arrangement of genes encoding immunoglobulins remains an important task, but these uncertainties may better be answered by molecular biological studies. Perhaps the more exciting avenues of research will involve the idiotype, not as a genetic marker, but rather as an essential recognitive physiological component which can function as a regulator of a complex dynamic system.

REFERENCES

1. Osterland, C.K., Miller, E.J., Karakawa, W.W. and Krause, R.M. (1966) J. Exp. Med. 123, 559-614.

2. Greenblatt, J.J., Bernstein, D., Bokisch, V.A., Kindt, T.J. and Krause, R.M. (1973) J. Immunol. 110, 862-866.

3. Kindt, T.J. (1975) Adv. Immunol. 21, 35-86.

4. Eichmann, K. and Kindt, T.J. (1971) J. Exp. Med. 134, 532-552.

5. Kindt, T.J., Seide, R.K., Bokisch, V.A. and Krause, R.M. (1973) J. Exp. Med. 138, 522-537.

6. Braun, D.G. and Kelus, A.S. (1973) J. Exp. Med. 138, 1248-1265.

7. Kindt, T.J. and Krause, R.M. (1974) Ann. Immunol. 125C, 369-372.

8. Sogn, J.A., Yarmush, M.L. and Kindt, T.J. (1976). Ann. Immunol. 127C, 397-408.

9. Yarmush, M.L., Sogn, J.A., Mudgett, M. and Kindt, T.J. (1977) J. Exp. Med. 145, 916-930.

10. Sogn, J.A., Yarmush, M.L. and Kindt, T.J. (1977) in Immune System: Genetics and Regulation, Sercarz, E.E., Herzenberg, L.A. and Fox, C.F. eds., Academic Press, New York, pp. 91-97.

11. Coligan, J.E., Schnute, W.C., Jr. and Kindt, T.J. (1975) J. Immunol. 114, 1654-1658.

12. Coligan, J.E., Fraser, B.A. and Kindt, T.J. (1977) J. Immunol. 118, 6-11.

13. Coligan, J.E., Fraser, B.A. and Kindt, T.J. (1978) in Cell Surface Carbohydrates and Biological Recognition, Marchesi, V.T., ed., Alan R. Liss, Inc., New York, pp. 601-612.

14. Krause, R.M. (1963) Bacteriol. Rev. 27, 369-380.

15. McCarty, M. (1971) Harvey Lect. 65, 73-96.

16. Coligan, J.E., Krause, R.M. and Kindt, T.J. (1978) Immunochemistry, (in press).

17. Krause, R.M. (1970) Adv. Immunol. 12, 1-56.

18. Aasted, B. and Kindt, T.J. (1976) Eur. J. Immunol. 6, 721-727.

19. Eichmann, K., Braun, D.G. and Krause, R.M. (1971) J. Exp. Med. 134, 48-65.

20. Greenblatt, J.J., Eichmann, K., Braun, D.G. and Krause, R.M. (1971) J. Infect. Dis. 124, 387-393.

21. Klapper, D.G. and Kindt, T.J. (1974) Scand. J. Immunol. 3, 483-490.

22. Winfield, J.B., Pincus, J.N. and Mage, R.G. (1972) J. Immunol. 108, 1278-1287.

23. Oudin, J. and Bordenave, G. (1971) Nature, (New Biol.) 231, 86-87.

24. Oudin, J. (1975) in The Antigens, Vol. 2, Sela, M. ed., Academic Press, New York, pp. 278-364.

25. Sogn, J.A., Coligan, J.E. and Kindt, T.J. (1977) Fed. Proc. 36, 214-220.

26. Oudin, J. and Michel, M. (1969) J. Exp. Med. 130, 595-617.

27. Oudin, J. and Michel, M. (1969) J. Exp. Med. 130, 619-642.

28. Eichmann, K., Braun, D.G., Feizi, T. and Krause, R.M. (1970) J. Exp. Med. 131, 1169-1189.

29. MacDonald, A.B. and Nisonoff, A. (1970) J. Exp. Med. 131, 583-601.

30. Brient, B.W. and Nisonoff, A. (1970) J. Exp. Med. 132, 951-962.

31. Claflin, J.L. and Davie, J.M. (1975) J. Immunol. 114, 70-75.

32. Mudgett, M. Coligan, J.E. and Kindt, T.J. (1978) J. Immunol. 120, 293-301.

33. Grey, H.M., Mannik, M. and Kunkel, H.G. (1965) J. Exp. Med. 121, 561-572.

34. Ghose, A.C. and Karush, F. (1974) J. Immunol. 113, 162-172.

35. Huser, H., Haimovich, J. and Jaton, J.C. (1975) Eur. J. Immunol. 5, 206-210.

36. Knight, K.L. and Hanly, W.C. (1975) in Contemporary Topics in Molecular Immunology, Vol. 4, Inman, F.P. and Mandy, W.J. eds., Plenum Publishing Corp., New York. pp. 55-88.

37. Yarmush, M.L., Mandy, W.J. and Kindt, T.J., Manuscript in preparation.

38. Kindt, T.J., Klapper, D.G. and Waterfield, M.D. (1973) J. Exp. Med. 137, 636-648.

39. Aasted, B. and Kindt, T.J. (1976) Eur. J. Immunol. 6, 727-732.

40. Thunberg, A.L. and Kindt, T.J. (1974) Eur. J. Immunol. 4, 478-483.

41. Capra, J.D. and Kehoe, J.M. (1974) Proc. Natl. Acad. Sci. 71, 4032-4036.

42. Fraser, B.A., Thunberg, A.L. and Kindt, T.J. (1978) Eur. J. Immunol. 8, 380-385.

43. Wu, T.T. and Kabat, E.A. (1970) J. Exp. Med. 132, 211-250.

44. Capra, J.D. and Kindt, T.J. (1975) Immunogenetics, 1, 417-427.

45. Kindt, T.J. and Capra, J.D. (1978) Immunogenetics, 6, 309-314.

46. Cazenave, P.A. (1977) Proc. Natl. Acad. Sci. 74, 5122-5125.

47. Urbain, J., Wikler, M., Franssen, J.D. and Collignon, C. (1977) Proc. Natl. Acad. Sci. 74, 5126-5130.

48. Jerne, N.K. (1976) Harvey Lect. 70, 93-110.

49. Rodkey, L.S. (1974) J. Exp. Med. 139, 712-720.

50. Cosenza, H., Julius, M. and Augustin, A. (1977) Immunol. Rev. 34, 3-33.

51. McKearn, T.J., Stuart, F.P. and Fitch, F.W. (1974) J. Immunol. 113, 1876-1882.

52. Eichmann, K. (1978) Adv. Immunol. 26, 195-254.

53. Ramseier, H., Aguet, M. and Lindenmann, J. (1977) Immunol. Rev. 34, 50-88.

54. Binz, H. and Wigzell, H. (1977) in Contemporary Topics in Immunobiology, 7, Stutman, O. ed., Plenum Press, New York 113-177.

55. Krammer, P.H. (1978) J. Exp. Med. 147, 25-38.

56. Nisonoff, A., Ju, S.-T. and Owen, F.L. (1977) Immunol. Rev. 34, 89-118.

57. Eardley, D., Shen, F.W., Cantor, H., and Gershon, R.K. (1977) in The Immune System: Genetics and Regulation, Sercarz, E.E., Herzenberg, L.A. and Fox, C.F., eds., Academic Press, New York, pp. 525-531.

58. Eichmann, K. (1977 in The Immune System: Genetics and Regulation, Sercarz, E.E., Herzenberg, L.A., and Fox, C. F., eds., Academic Press, New York, pp. 127-138.

59. Eichmann, K. and Rajewsky, K. (1974) Eur. J. Immunol. 5 661-666.

60. Oudin, J. and Cazenave, P.A. (1971) Proc. Natl. Acad. Sci. 68, 2616-2620.

61. Urbain, J., Tasiaux, N., Leuwenkroon, R., Van Acker, A. and Mariame, B. (1975) Eur. J. Immunol. 5, 570-575.

Rudbach/Baker, eds. Immunology of Bacterial Polysaccharides

REGULATION OF THE ANTIBODY RESPONSE TO PNEUMOCOCCAL POLYSACCHARIDES BY THYMUS-DERIVED (T) CELLS: MODE OF ACTION OF SUPPRESSOR AND AMPLIFIER T CELLS

PHILLIP J. BAKER AND BENJAMIN PRESCOTT
Laboratory of Microbial Immunity, National Institute of Allergy and Infectious Diseases, National Institutes of Health, Bethesda, Maryland 20014

ABSTRACT

The results of detailed studies on the mechanisms by which thymus-derived (T) amplifier and suppressor cells influence the magnitude of the antibody response to Type III pneumococcal polysaccharide (SSS-III) are reviewed. Amplifier T cells, which are functionally distinct from helper T cells, act mainly to drive antigen-stimulated bone marrow-derived precursors of antibody-forming cells (B cells) to proliferate further in response to antigen. By contrast, suppressor T cells function in at least two different ways; by interferring, perhaps in a competitive manner, with the expression of amplifier T cell activity, and by acting directly on B cells to limit the extent to which they proliferate and/or differentiate after stimulation by antigen. The nature of the recognition units involved in promoting these cellular interactions is discussed along with some of the genetic data obtained.

INTRODUCTION

It has been established that athymic nude mice, as well as thymus-deprived adult mice, give a normal antibody response to

Type III pneumococcal polysaccharide (SSS-III), a linear polymer composed solely of glucose-glucuronic acid disaccharide subunits; consequently, the antibody response to this well-defined bacterial antigen has long been considered to be thymus-independent [1-6]. However, the results of other studies indicate that the antibody response to SSS-III is regulated, nevertheless, by at least two different types of thymus-derived (T) cells having opposing functions. Such regulatory T cells have been referred to as amplifier and suppressor T cells [5,7].

As a result of extensive studies conducted in our laboratory during the past ten or more years, the antibody response to SSS-III has been well-characterized at the cellular level, and many of the properties of regulatory T cells have been defined [7]. The main purpose of this brief report is to summarize some of our published observations in this regard which, when considered together, best describe how suppressor and amplifier T cells act to regulate the magnitude of an antibody response. These findings illustrate that the antibody response to SSS-III provides not only an ideal experimental model for investigating the cellular interactions involved in regulating antibody responses in general, but also a means for evaluating the biomedical and genetic implications of such phenomena to the development of effective immunity against infectious agents.

MATERIALS AND METHODS

Animals. BALB/c female mice (8-12 weeks of age), obtained from the Cumberland View Farms, Clinton, Tennessee, were used for most of this work. Recombinant-inbred (RI) strains of mice were purchased from the Jackson Laboratories, Bar Harbor, Maine.

Antigens. The immunologic properties of the Type III pneumococcal polysaccharide (SSS-III) used, as well as the method by which it was prepared have been described in detail [7-10]. For immunization, mice were given a single i.p. injection of an optimally immunogenic dose (0.5 μg) of SSS-III in 0.5 ml normal saline.

Trinitrophenyl (TNP)-SSS-III was prepared by first synthesizing an amino derivative of SSS-III using a modification of the method of Jones *et al.*, [11]. Here, 50 mg of SSS-III (10mg/ml in borate saline buffer, pH 8.0) was placed in a beaker kept in an ice bath. Then 0.67 ml (0.1 moles) of anhydrous ethylene diamine, and 0.250 ml of 0.1 M sodium metaperiodate were added with constant stirring. After one minute, 1.25 ml of 2.0 M sodium borohydride was added and the mixture was stirred for one hour; at the end of this time period, the solution was dialyzed extensively against several changes of borate saline buffer. After dialysis, 1.0 μM of 2, 4, 6 - trinitrobenzene sulfonic acid, in a volume of 0.2 ml borate saline buffer, was added to a 2.0 ml portion of the amino derivative made (2 mg/ml); the mixture was stirred at room temperature for one hour, after which it was dialyzed against several changes of borate saline buffer. The number of TNP groups per mole of SSS-III (average molecular weight 100,000) was determined spectrophotometrically [12] and found to be 7 TNP groups per mole SSS-III (TNP_7-SSS-III). For immunization, mice were given a single i.p. injection of 0.5 μg or 0.05 μg of TNP_7-SSS-III in 0.5 ml normal saline.

Dextran B-1355 (Lot A), a branched glucose polymer derived from Leuconostoc mesenteroides (NRRL-B-1355), was prepared in our laboratory by the method of Jeanes et al.[13]. For immunization, mice were given a single i.p. injection of an optimally immunogenic dose (100 μg) in 0.2 ml normal saline.

Immunological Methods. Numbers of antibody-producing plaque-forming cells (PFC), specific for each of the antigens used in this work, provided a measure of the magnitude of the antibody response produced, 5 days after immunization. In this procedure, each plaque or area of complete lysis represents a single cell synthesizing and secreting antibody specific for the immunizing antigen. PFC making antibody of the IgM class were detected by a slide version of the technique of localized hemolysis-in-gel [14] using indicator cells sensitized with SSS-III [16], TNP [12], or dextran [15] by conventional methods. Only PFC specific for the immunizing antigen are considered; the results obtained, which are log-normally distributed, are expressed as the geometric mean of the $\log_{10}$ number of PFC/spleen for groups of similarly treated mice. Student's t test was used to assess the significance of the differences observed; differences were considered to be significant when probability (p) values of <0.05 were obtained.

Other Materials. Horse anti-mouse lymphocyte serum (ALS), Lot 13162, was purchased from Microbiological Associates, Inc., Bethesda, Maryland. Velban (vinblastine sulfate), a mitotic inhibitor, was purchased from Eli Lilly and Company, Indianapolis, Indiana. Concanavalin A (Con A), carbohydrate content <0.1%, was purchased from Pharmacia Fine Chemicals, Piscataway, New Jersey. Pertinent references concerning the immunological properties of these materials, as well as the manner in which they were employed, are given in the text.

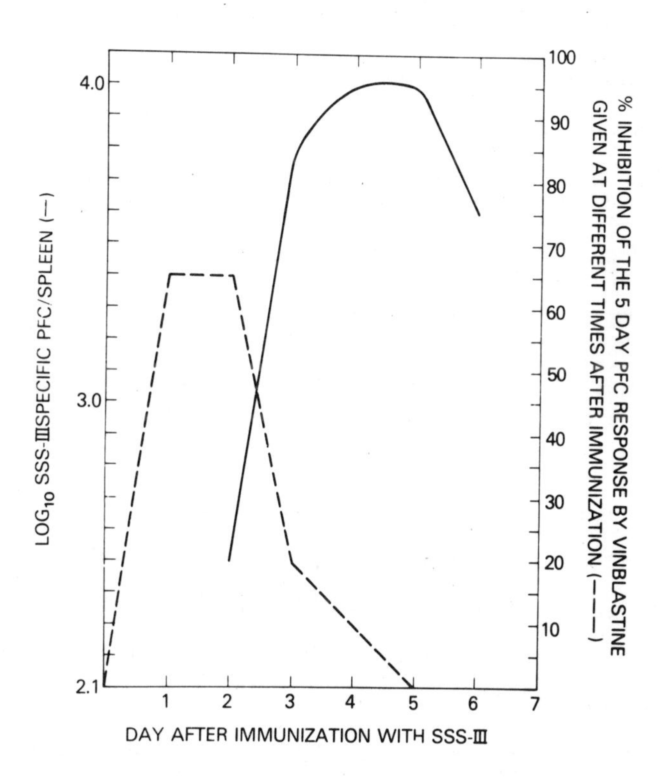

Fig. 1. Effect of Velban (vinblastine sulfate) on the magnitude of the 5 day PFC response of mice to an optimally immunogenic dose (0.5 μg) of SSS-III. A single injection of Velban (50 μg) was given (i.p.) on different days relative to immunization; the inhibitory effect of such treatment on the number of PFC detected 5 days after immunization was determined (broken line). Numbers of PFC detected on different days after immunization with SSS-III for mice not given Velban are also given (solid line). All of the data shown are taken from reference 16.

RESULTS AND DISCUSSION

The data of Figure 1 show, in a diagramatic manner, some general features associated with the antibody response of mice to SSS-III. If a mitotic inhibitor, such as Velban (vinblastine sulfate), is given (50 µg, i.p.) at different times after immunization with an optimally immunogenic dose (0.5 µg) of SSS-III, a significant reduction in the magnitude of the 5 day (peak) PFC response occurs only when the drug is given during the first 24-48 hours after immunization [16]; note that this is well before significant numbers of SSS-III-specific PFC can be detected [16]. This means that bone marrow-derived precursors of antibody-forming cells (B cells) must be proliferating during this time interval, and the extent to which such cells proliferate determines the number of SSS-III-specific PFC eventually produced by day 5. These findings are in complete agreement with the results of earlier studies showing that antigen-binding B cells, specific for pneumococcal polysaccharides, begin to increase almost immediately after immunization and reach maximal numbers by 48-72 hours after immunization with an optimally immunogenic dose of antigen [17,18]. By contrast, the data of Figure 1 also show that SSS-III-specific PFC start to appear in significant numbers <u>after</u> 48 hours and reach maximal numbers by 4-5 days after immunization [16]. Thus, virtually all of the proliferation that takes place during the course of an antibody response to SSS-III occurs during the first 2 days after immunization, before the appearance of PFC; this proliferation involves mainly antigen-binding B cells or precursors of PFC. It should be noted that the same or larger doses of Velban, given on day 4, do not reduce the numbers of SSS-III-specific PFC detected on day 5; this indicates that, under the test conditions employed, the drug did not interfere with the synthesis and release of antibody by individual PFC [16].

What happens to this kinetic pattern when suppressor T cells, which are activated during the course of a normal antibody response to SSS-III, are eliminated by treatment with a T cell depleting agent such as anti-lymphocyte serum or ALS? The data of Figure 2 show that during the first 3 days after immunization, the kinetics for the appearance of SSS-III-specific PFC are similar for both ALS-treated and non-ALS-treated mice. However, in the case of ALS-treated mice, PFC continue to increase at an exponential rate beyond day 3; this results in a 20-fold increase in the magnitude of the PFC response [16]. Such ALS-induced enhancement can be eliminated completely by the administration of a mitotic inhibitor (vinblastine sulfate) on day 4 [16]; therefore, this enhancement is due mainly to additional rounds of cellular proliferation. Consequently, the removal of suppressor T cell activity by treatment with ALS results in further rounds of proliferation on the part of antigen-stimulated B cells. It should be noted that maximal enhancement is obtained _only_ when ALS is given on the day of immunization with SSS-III (day 0); no enhancement occurs when ALS is given 2 or 3 days later [16]. This means that suppressor T cells are activated _early_ during the course of an immune response, perhaps as early as 18-24 hours after immunization, and they begin to exert their inhibitory effects _before_ the appearance of significant numbers of PFC.

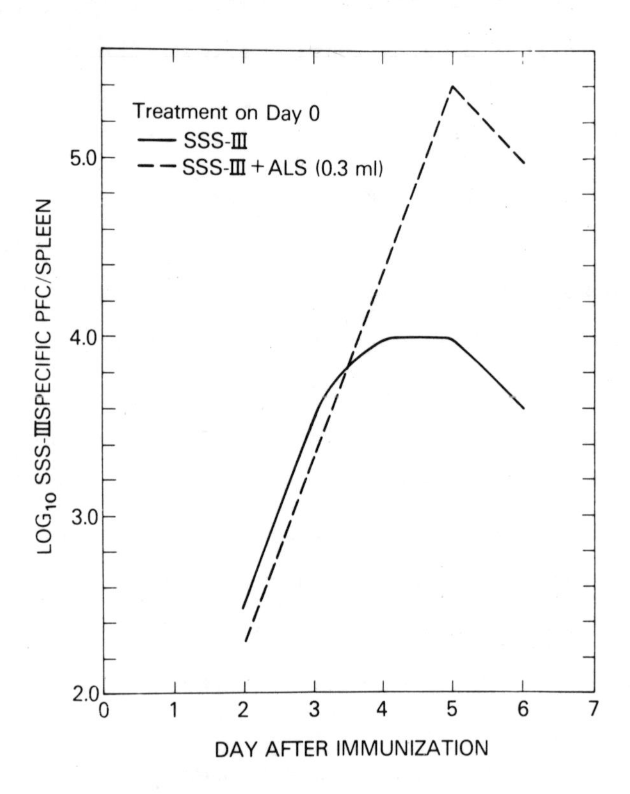

Fig. 2. Effect of treatment with ALS on the magnitude of the PFC response of mice to an optimally immunogenic dose (0.5 μg) of SSS-III. ALS (0.3 ml) was given (i.p.) at the time of immunization with SSS-III. The data shown are taken from reference 16.

Since athymic nude mice, as well as thymus-deprived adult mice, give a normal, rather than enhanced, antibody response to SSS-III [1-5], ALS-induced enhancement cannot be attributed solely to the removal of inhibitory suppressor T cells; instead, it is a T cell dependent process that requires the participation of another subpopulation of T cells. We have used the term <u>amplifier T cells</u> to designate these cells, which have a positive influence on the magnitude of the antibody response [5,7]. If a T cell stimulator such as Concanavalin A (Con A) is given 2 days after immunization with SSS-III, a significant ($p < 0.05$) increase in the PFC response occurs (Table 1); such enhancement, like the enhancement produced after treatment with ALS, (a) cannot be demonstrated in athymic nude mice, (b) is T cell dependent, and (c) results in the further proliferation of antigen-stimulated B cells [19]. From these and the kinetic studies summarized above, one may conclude that amplifier T cell activity begins to be expressed between 2 and 3 days following immunization, i.e., after suppressor T cells have been activiated, and that the major effect produced by amplifier T cells is to cause antigen-stimulated B cells to undergo further rounds of proliferation [19]. Since this additional proliferation occurs only when suppressor T cell activity is eliminated and does not occur in the absence of amplifier T cells, the main effect produced by suppressor T cells, during the course of a normal response to an optimally immunogenic dose of SSS-III, is to interfere with the expression of amplifier T cell activity [19]. The net result of such inhibition is to limit--in an indirect manner -- the extent to which populations of B cells expand or increase in size in response to antigen. Such a scheme provides a satisfactory explanation for the fact that the antibody response of athymic nude mice to SSS-III is essentially the same in magnitude as that produced by intact normal mice having both types of regulatory T cells; in both cases, the effects produced by amplifier T cells are either absent, or not expressed.

TABLE 1

ABILITY OF CONCANAVALIN A (CON A) TO INCREASE THE PFC RESPONSE OF MICE TO AN OPTIMALLY IMMUNOGENIC DOSE (0.5 μg) OF SSS-III

Treatment [a]		
Con A	SSS-III	Log_{10} PFC/spleen [b]
--	0.5 μg	4.108 ± 0.126 (12,811)
300 μg	0.5 μg	4.976 ± 0.090 (94,599)

[a] Con A was given, 2 days after immunization with SSS-III. Both Con A and SSS-III were given intraperitoneally.

[b] Mean ± S.E.M. for groups of 5 mice, 5 days after immunization with SSS-III; geometric means are in parentheses.

Because of the way in which the preceeding experiments were conducted, one could not determine whether amplifier T cells are antigen-specific in their mode of action. However, the results of a study summarized in Table 2 imply -- but do not necessarily prove -- that this _might_ be the case. In this particular study, mice were given ALS at the time of immunization with TNP_7-SSS-III; numbers of both SSS-III specific and TNP-specific PFC were determined, 5 days later and the results obtained were compared to those for non-ALS-treated mice. The data show that enhancement (about 10 fold) was evident only with respect to the SSS-III-specific PFC response; no enhancement was noted with respect to the magnitude of the TNP-specific PFC response. These data refute the argument that ALS-induced enhancement is due simply to splenomegaly or to a non-specific mitogenic effect of ALS on B cells; if that were the case, PFC responses to _both_ SSS-III and TNP would have been increased to about the same extent.

TABLE 2

EFFECT OF TREATMENT WITH ANTILYMPHOCYTE SERUM (ALS) ON THE PFC RESPONSE TO BOTH SSS-III AND TNP IN MICE IMMUNIZED WITH TNP_7-SSS-III

Treatment [b]	Log_{10} PFC/spleen vs. [a]	
	SSS-III	TNP
0.5 μg TNP_7-SSS-III	4.066 ± 0.108 (11,635)	2.896 ± 0.069 (786)
0.5 μg TNP_7-SSS-III + ALS	5.035 ± 0.095 (108,434)	2.763 ± 0.327 (778)
None	None detected	(<50)

[a] Mean ± S.E.M. for groups of 5-8 mice, 5 days after immunization; geometric means are in parentheses.

[b] ALS (0.3 ml) was given intraperitoneally at the time of immunization.

Rather, these findings suggest that amplifier T cells involved in regulating the antibody response to SSS-III may not influence antibody responses to other unrelated antigens. In another experiment (data not shown), we found that treatment with ALS does not increase the magnitude of the PFC response of BALB/c mice to an optimally immunogenic dose of dextran B-1355; however, when BALB/c mice are given ALS at the time of immunization with <u>both</u> SSS-III and dextran B-1355, the PFC response to SSS-III is increased by more than 15 fold whereas the PFC response to dextran is unchanged.

There are several observations which indicate that amplifier T cells and helper T cells represent functionally distinct populations of lymphocytes. First, helper T cells -- perhaps by means of an antigen-presentation mechanism -- act at an early stage of the antibody response (<24 hr after immunization) and thus appear to be needed for the initiation of antibody formation to some antigens; by contrast, the results of our studies, which have been reviewed above, first of all clearly show that amplifier T cells act at a later stage of the antibody response (2-3 days after immunization) and drive antigen-stimulated B cells to proliferate further in response to antigen [19]. Second, helper T cell activity is either completely eliminated or substantially reduced within 28 weeks after adult thymectomy and by treatment with ALS at the time of immunization with helper T cell dependent antigens; amplifier T cell activity is not impaired in the least by any of these experimental procedures [20]. Third, the activity of helper T cells appears to be governed by genes within the major histocompatibility (H-2) complex; however, the results of genetic studies conducted in our laboratory reveal that amplifier T cell activity is not H-2 associated [21,22]. Fourth, both the induction and expression of helper T cell activity can be eliminated by treatment with hydrocortisone succinate; this steroid has no effect on either the induction or expression of amplifier T cell activity [23]. Since functional differences between helper and amplifier T cells have been described also by other investigators [24], it is clear that such cells must represent distinct subpopulations of T cells whose activities must be examined and analyzed separately.

Although ALS-induced enhancement suggests that suppressor T cells, but not amplifier T cells, are susceptible to inactivation by ALS, this is not really the case. Since treatment with ALS,

2-3 days before immunization with SSS-III has no effect on the magnitude of the PFC response produced [16], one may assume that precursors of both suppressor and amplifier T cells are relatively resistant to inactivation by ALS. Consequently, differences in the times at which these cells become activated after immunization must account for their apparent differential sensitivity to ALS [25]. From information available on the degree of ALS-induced enhancement obtained when ALS is given on different days before immunization with SSS-III [16], one may construct a curve showing the decay or decline of enhancing activity with time after the administration of ALS (Figure 3). Also shown in Figure 3 are curves describing the kinetics for the appearance of both suppressor and amplifier T cell activity with time after immunization with an optimally immunogenic dose of SSS-III; the latter are only approximations based on the best available experimental data for the times at which these activities are first detectable and reach maximal levels after immunization [11, 16, 19, 26]. It is clear that if a single injection of ALS is given at the time of immunization with SSS-III (day 0), sufficient amounts of ALS remain in the circulation during the first 2 days to eliminate suppressor T cell activity, almost as soon as it begins to be expressed. However, amplifier T cell activity escapes inactivation, since this activity develops later and at a time when most of the ALS given on day 0 has been cleared or eliminated from the circulation. According to this scheme, one would predict that if a second injection of ALS were given on day 2, amplifier T cell activity likewise would be eliminated and no ALS-induced enhancement would be obtained. This is exactly what happens;

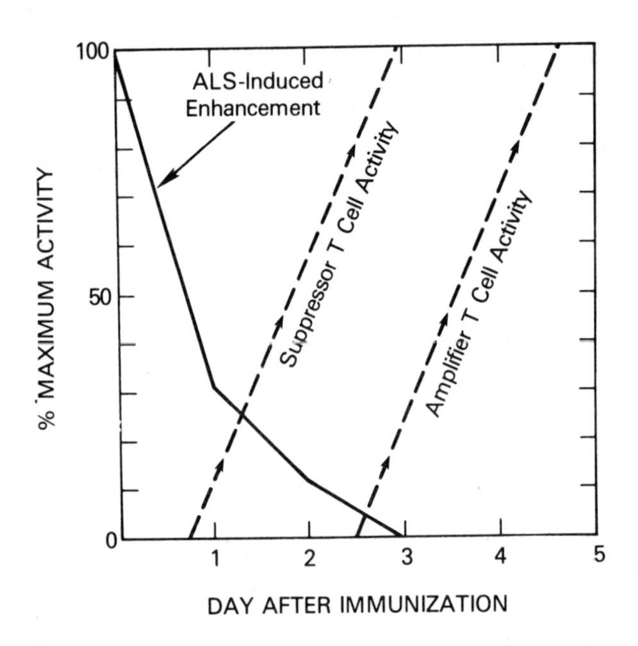

Fig. 3. Kinetic relationships between the decline of circulating ALS, with respect to its enhancing activity (solid line) and the rate of appearance of both suppressor T cell and amplifier T cell activity for mice given an optimally immunogenic dose (0.5 µg) of SSS-III (broken lines). The curves shown are based on values given in references 11, 16, 19, and 26.

one gets a normal, rather than an enhanced, antibody response [25]. Since two injections of ALS, given on day 0, do not alter the degree of enhancement obtained [25], this means that suppressor and amplifier T cells --once activated-- are indeed ALS-sensitive and that the basis for the apparent differential sensitivity noted in our earlier studies [27] is due mostly to differences in the times at which these cells become activated after immunization, in relation to the amount of ALS present in the circulation at such time intervals [25]. Obviously, a consideration of such kinetic relationships is important if one wishes to understand the mode of action of suppressor and amplifier T cells in any given experimental situation.

Up to now, suppressor T cells have been discussed only with respect to their inhibitory effects on the expression of amplifier T cell activity. However, this is not the only way in which suppressor T cells function; under some experimental conditions, they can also act directly on B cells to limit the extent to which B cells proliferate in response to antigen. Perhaps the best way to illustrate how this occurs is to examine the phenomenon of <u>low-dose paralysis</u> [10, 28], which is described by the data of Table 3. Here it is shown that pretreatment or priming with an extremely low or marginally immunogenic dose (0.005 µg) of SSS-III greatly reduces the magnitude of a PFC response to an optimally immunogenic dose (0.5 µg) of SSS-III given 3 or more days later; the term low-dose paralysis has been used to describe this particular form of unresponsiveness which has been induced in every strain of mice examined thus far, regardless of major (H-2) histocompatibility type, and persists for at least 2 1/2 months after priming with a single injection of antigen [10, 20]. The results of several studies have shown that low-dose paralysis is an antigen-specific, T cell dependent phenomenon, that can be induced under conditions in which priming does not lead to the formation of detectable serum antibody or PFC [10, 28, 29]. The ability of T cell depleting agents (ALS or lactic dehydrogenase virus) to abrogate this form

TABLE 3

EFFECT OF PRIOR TREATMENT WITH A LOW DOSE (0.005 μg) OF SSS-III ON THE MAGNITUDE OF THE PFC RESPONSE TO AN OPTIMAL DOSE (0.5 μg) OF SSS-III

Priming Dose [a]	Immunizing Dose	Log_{10} PFC/spleen [b]
--	0.5 μg	4.293 ± 0.041 (19,636)
0.005 μg	0.5 μg	3.541 ± 0.095 (2,541)

[a] Mice were primed with 0.005 μg of SSS-III, 3 days before immunization with 0.5 μg of SSS-III.

[b] Mean ± S.E.M. for groups of 10 mice, 5 days after immunization; geometric means are in parentheses.

of unresponsiveness [28, 30], enabling mice to give once more a normal antibody response, clearly indicates that low-dose paralysis results in active supression --rather than the deletion or irreversible inactivation-- of B cells capable of responding to SSS-III [28]. The specificity of the expression of low-dose paralysis can best be illustrated by examining the results of two separate experiments, summarized in Tables 4 and 5.

In the first experiment (Table 4), mice were primed with a low dose (0.005 μg) of SSS-III and then they were immunized, 3 days later, with an optimally immunogenic dose of both SSS-III (0.5 μg) and dextran B-1355 (100 μg). PFC specific for both antigens were determined, 5 days after immunization and the results obtained were compared to those for unprimed mice given both antigens together. The data (Table 4) show that priming with a

TABLE 4

EFFECT OF PRIMING WITH A LOW DOSE (0.005 μg) OF SSS-III ON THE MAGNITUDE OF THE PFC RESPONSE TO BOTH SSS-III AND DEXTRAN B-1355

Treatment		Log_{10} PFC/spleen [a] vs.	
Priming[b]	Immunization	SSS-III	DEXTRAN
--	0.5 μg SSS-III + 100 μg Dextran	4.150 ± 0.060 (14,127)	4.601 ± 0.053 (39,888)
0.005 μg SSS-III	0.5 μg SSS-III + 100 μg Dextran	3.182 ± 0.132 (1,520)	4.608 ± 0.080 (40,554)

[a] Mean ± S.E.M. for groups of 10 mice, 5 days after immunization with both SSS-III and dextran; geometric means are in parentheses.

[b] Mice were primed (pre-treated) with a low dose (0.005 μg) of SSS-III, 3 days before immunization with both SSS-III and dextran. Both antigens were given intraperitoneally.

low dose of SSS-III caused significant suppression ($p < 0.005$) of the PFC response to SSS-III; by contrast, the PFC response to dextran, which does not crossreact with SSS-III, was not influenced by priming with SSS-III ($p > 0.05$). In the second experiment (Table 5), mice were primed with a low dose (0.005 μg) of SSS-III and then they were immunized, 3 days later, with 0.05 μg of TNP_7-SSS-III; numbers of both SSS-III-specific and TNP-specific PFC were determined 5 days after immunization and the results obtained were compared to those for unprimed mice. The data (Table 5) show that priming with SSS-III results in significant suppression ($p < 0.005$) of <u>only</u> the SSS-III-specific PFC response;

TABLE 5

EFFECT OF PRIMING WITH A LOW DOSE (0.005 μg) OF SSS-III ON THE PFC RESPONSE TO BOTH SSS-III AND TNP IN MICE IMMUNIZED WITH TNP_7-SSS-III

Treatment		Log_{10} PFC/spleen [a] vs.	
Priming [b]	Immunization	SSS-III	TNP
--	0.05 μg TNP_7-SSS-III	3.522 ± 0.163 (3,327)	2.587 ± 0.091 (384)
0.005 μg SSS-III	0.05 μg TNP_7-SSS-III	2.614 ± 0.088 (411)	2.633 ± 0.050 (430)

[a] Mean ± S.E.M. for groups of 10 mice, 5 days after immunization with 0.05 μg TNP_7-SSS-III; geometric means are in parentheses.

[b] Mice were primed with 0.005 μg of SSS-III, 3 days before immunization with TNP_7-SSS-III.

the TNP-specific response was not affected ($p > 0.05$). The results of these two experiments establish that, unlike the phenomenon of competition between antigens, the phenomenon of low-dose paralysis is specific with respect to both its induction and expression.

There are several reports attesting to the fact the Concanavalin A (Con A), a T cell mitogen, can non-specifically suppress a variety of immune responses *in vitro*; such effects have been attributed to soluble products released from mitogen-activated suppressor T cells [31, 32]. These reports prompted us to investigate whether the *in vivo* administration of Con A would result in suppression of the PFC response to SSS-III and whether the kinetics for the development of such suppression would resemble those seen after the induction of low-dose paralysis;

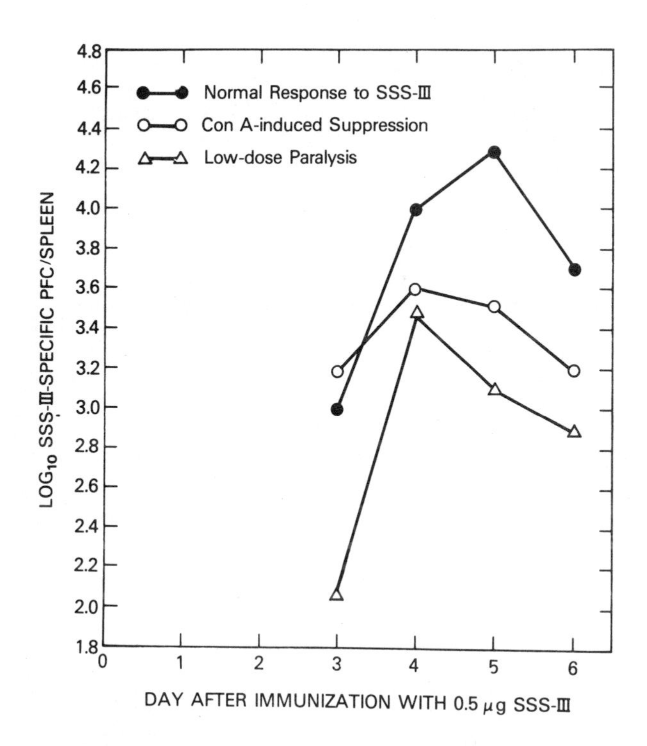

Fig. 4. Numbers of PFC detected after immunization of mice with an optimally immunogenic dose (0.5 μg) of SSS-III (closed circles), and after the induction of low-dose paralysis (triangles) or Con A-induced suppression (open circles). The kinetic data shown are taken from reference 33.

this was indeed found to be the case [33], as shown by the data of Figure 4. Con A given at the time of immunization with an optimally immunogenic dose (0.5 μg) of SSS-III induced suppression of the PFC response which was at least as great (>80%) as that found after the induction of low-dose paralysis. As was the case for low-dose paralysis, Con A-induced suppression cannot be induced in athymic nude mice and could be eliminated by treatment with T cell depleting agents [33]; this affirms that it is a T cell dependent phenomenon [33]. More important, the kinetics for the development of Con A-induced suppression parallel those for the expression of low-dose paralysis, suggesting that --except for specificity-- both forms of suppression share a common pathway by which suppression is mediated [33]. Since Con A is able to induce suppression only when given at the time of immunization [33], and since the induction of low-dose paralysis requires a latent or inductive period of 2-3 days for maximal expression [10, 16, 28], one may conclude that, under these experimental conditions, suppressor T cells act at an early stage of the immune response, i.e., before the appearance of PFC. Because suppressor T cells have been activated or primed before the second encounter with antigen in the low-dose paralysis situation, they most likely act directly on B cells to limit the extent to which precursors of PFC proliferate and/or differentiate in response to an optimally immunogenic dose of antigen [33].

The results of these studies have provided much information on the mechanisms by which amplifier and suppressor T cells influence the magnitude of an antibody response. But, the recognition units involved in the activation of these regulatory T cells remain to be characterized. Since the inhibitory effects produced by suppressor T cells are antigen-specific in the low-dose paralysis situation (Table 3; reference 28), the simplest explanation is that suppressor--and perhaps amplifier-- T cells recognize and respond directly to antigen itself; in other words, these cells possess some sort of antigen-specific receptor on their surface. However, the inability of SSS-III to elicit T cell cooperative effects, immunologic memory, delayed hypersensitivity reactions, and T cell proliferative responses suggest --but does not necessarily prove-- that T cells may not be able to recognize this particular antigen [1, 2, 34-36]. Furthermore, the suppression produced after the induction of low-dose paralysis is similar in both degree and kinetics to that found after treatment with Con A, a non-specific stimulator of T cells (Figure 4; reference 33). This implies that the product made by suppressor T cells may not be antigen-specific; but, the recognition units involved in promoting the interactions between suppressor T cells and other cells may permit this non-specific product, which may act only over a very short range, to become focused upon a particular set or clone of cells. Such a focusing mechanism would permit a non-specific product made in extremely small amounts during the course of a normal immune response to act with a high degree of efficiency and in a manner that appears to be antigen-specific [7]. If regulatory T cells are unable to recognize antigens such as SSS-III, then what do they recognize? Actually, there is no satisfactory answer to this question at the present time. One possiblity is that both suppressor and amplifier T cells become activated, not in response to antigen per se, but in response to the idiotypic or unique determinant of antibody specific for SSS-III; as has been mentioned earlier, such determinants are normally present on the surface of B cells capable of responding to SSS-III and they increase in density per cell almost immediately after immunization [7, 19], i.e., prior to the activation of suppressor T cells and the appearance of PFC. According to this scheme, neither type of regulatory T cell would respond directly to antigen; instead, they would respond to or recognize increased or threshold amounts of cell-associated antibody on the surface of antigen-stimulated B cells, enabling them to interact with particular clones of B cells in a selective and perhaps competitive manner. The results of some of our most recent genetic studies using recombinant-inbred strains of mice that differ greatly in their ability to respond to SSS-III suggest that such competitive interactions between suppressor and amplifier T cells may in fact occur.

Even though the antibody response to SSS-III is highly restricted with respect to the immunoglobulin class [8] and the avidity [37] of the antibody produced, the genetic mechanisms that govern immunological responsiveness to this antigen appear to be quite complex. They include a gene(s) linked to the X chromosome that determines responsiveness in an almost "all-or-none" manner, as well as autosomal genes that influence the magnitude of the antibody response of mice possessing the X-linked gene [38, 39]. The effects governed by this second category of "quantitative genes" include the activities of B cells, amplifier T cells, and suppressor T cells, which undoubtedly are under separate genetic control [21, 22]. The combined activities of all three types of cells, therefore, determine the magnitude of the antibody response produced after immunization.

The availability of recombinant-inbred (RI) strains of mice, which are produced by inbreeding F_2 progeny from a cross between two genetically distinct progenitor strains [40, 41], enables one to investigate systematically the genetic control of B cell and regulatory T cell function with respect to the antibody response to SSS-III. In this context, the use of RI strains has a number of advantages. First, if the progenitor strains of mice differ greatly in responsiveness to an antigen, RI strains derived from such mice can be used to estimate the minimal number of genes involved in the expression of a given immunological trait. If all RI strains give an immune response identical to that of either progenitor strain in phenotype, a single gene could be involved. However, the occurrence of unique phenotypes among RI strains, i.e., phenotypes different from those of both progenitor strains, suggests the presence of new gene combinations (multigenic control); the minimal number of genes involved can be estimated from the frequency of unique phenotypes observed, since all genes present in both progenitor strains are conserved in the RI strains obtained. Second, some RI strains of mice have been well-characterized as to alleles present at a number of different genetic loci [42, 43]; this provides a convenient means for determining linkage between genes involved in the expression of any given trait and alleles present at other known loci.

In our genetic studies, which have been described in greater detail elsewhere [21, 22], RI strains of mice, as well as the progenitor strains from which they were derived, were examined with respect to three factors that contribute to the development of an antibody response to SSS-III. First, the magnitude of the PFC response to an optimally immunogenic dose of SSS-III was determined in order to evaluate the capacity of B cells to respond to this antigen; this provides a reasonably good index of intrinsic B cell activity, since the magnitude of the antibody response to SSS-III in athymic nude and thymus-deprived mice is essentially the same in magnitude as that of normal intact mice [1-5]. Second, the amount of enhancement produced after immunization and treatment with ALS was used to assess the influence of amplifer T cells on antigen-stimulated B cells in the absence of suppressor T cell activity; such enhancement has been found to be T cell dependent and not due simply to the inactivation of suppressor T cells [5, 7] or to a mitogenic effect of ALS on B cells (Table 2). Third, the degree of low-dose paralysis induced was used to determine the amount of suppressor T cell activity present; this form of unresponsiveness is an antigen-specific (Table 5), T cell dependent phenomenon in which suppressor T cells play an active role [5, 33]. When these same criteria were applied in other studies, each of the functional activities mentioned was found to mature independently and at different rates with age [20]. This experimental approach, therefore, enables one to assess the contribution of each functional activity separately and in such a way that the expression of one activity is not influenced, to a significant degree, by the remaining activities considered. All RI strains of mice, as well as BALB/cBy (C) and C57BL/6By (B) progenitor strains from which they were derived [40, 41], were purchased from the Jackson Laboratories, Bar Harbor, Maine. RI strains were designated as CXBD, CXBE, CXBG, CXBH, CXBI, CXBJ, and CXBK according to existing terminology [42, 43]; information concerning their H-2 type and IgC_H allotype, relative to that of progenitor C or B mice, has been reported [42, 43]. Only age-matched female mice were used to minimize the effects of environmental factors and the contribution of non-autosomal genes.

The data of Table 6 summarize the results obtained when RI and progenitor strains of mice were examined for their ability to respond to an optimally immunogenic dose (0.5 μg) of SSS-III; this provides a measure of the capacity of B cells to respond to SSS-III. The data make several significant points. First, 5 of 7 RI strains examined gave PFC responses that were unique, i.e., the mean values obtained differed significantly in magnitude from those of <u>both</u> progenitor B and C mice. All of these responses were greater than the response of low-responding B mice; however, the response of CXBG mice was <u>greater</u> than that of high-responding progenitor C mice, suggesting overdominance or complementarity between genes controlling this functional activity. Although the frequency of unique phenotypes found suggests that at least 5 autosomal genes may govern the capacity of B cells to respond to SSS-III, the responses of CXBJ and CXBI mice do not differ significantly ($p > 0.05$), and the responses of CXBD and CXBK mice do not differ from each other ($p > 0.05$). If one assumes that these pairs share the <u>same</u> new combination of genes, then a minimum of 3 -- rather than 5 -- autosomal genes are involved in determining responsiveness; however, this does not appear to be the case, since F_1 progeny obtained from crosses between these strains and low-responding B mice do not give the same responses (data not shown). Obviously, more than one autosomal gene must influence the capacity of B cells to respond to SSS-III. Since the recombination frequency observed

TABLE 6

PFC RESPONSE OF RI AND PROGENITOR B AND C MICE TO AN OPTIMALLY IMMUNOGENIC DOSE (0.5 μg) OF SSS-III

Mouse Strain	PFC response to SSS-III: PFC/spleen [a]	PFC response to SSS-III: Phenotype [b]	H-2 type	IgC_H allotype
B	2.485±0.095 (305)	B	B	B
CXBJ	2.921±0.137 (883)	Unique	B	C
CXBI	3.006±0.191 (1,014)	Unique	B	B
CXBD	3.723±0.075 (5,283)	Unique	C	B
CXBK	3.801±0.073 (6,320)	Unique	B	B
CXBH	3.975±0.091 (9,451)	C	C	B
C	4.102±0.078 (12,654)	C	C	C
CXBE	4.222±0.131 (16,679)	C	B	B
CXBG	4.306±0.039 (20,220)	Unique	B	C

[a] Log_{10} PFC/spleen ± S.E.M. for 10-16 mice, 5 days after immunization with 0.5 μg SSS-III; geometric means are in parentheses.

[b] B and C indicate that the response produced was like ($p > 0.05$) that of B or C mice, respectively; unique responses were significantly different ($p < 0.05$) from those of both B and C mice.

is > 0.5, one must consider the possibility that the genes involved may not even be on the same chromosome. Second, these data, as well as those of other studies [38, 44], confirm that the ability of mice to respond to SSS-III is not associated with immune response (Ir) genes, linked to either the H-2 complex or the IgC_H allotype complex.

The data of Table 7 show once more that B mice and CXBI mice both give low PFC responses to SSS-III; however, mice derived from a cross between these two low-responding strains give a PFC response, 7-20 times greater than that of either parental strain. Likewise, mice derived from a cross between B and CXBJ mice give a response 2-6 times greater than that of either parental strain. Thus, a cross between two low responding strains can give rise to progeny that are high to intermediate responders to SSS-III. Mice from a cross between high-responding C mice and low-responding CXBJ mice produce a PFC response like that of C mice ($p > 0.05$); high responsiveness, rather than low responsiveness, is dominant. These data indicate that genes governing the capacity of B cells to respond to SSS-III act in a complementary and dominant fashion [21].

The degree of unresponsiveness produced after pretreatment (priming) with a marginally immunogenic dose (0.005 μg) of SSS-III (low-dose paralysis) was used to determine the amount of suppressor T cell activity present in RI and progenitor strains of mice.

TABLE 7

COMPLEMENTARY EFFECTS OF GENES GOVERNING THE CAPACITY OF B CELLS TO RESPOND TO AN OPTIMALLY IMMUNOGENIC DOSE (0.5 μg) OF SSS-III

Strain	PFC/spleen [a]
B	2.485 $\pm$ 0.095 (305)
CXBI	3.006 $\pm$ 0.191 (1,014)
[(CXBI) x B] F_1	3.870 $\pm$ 0.081 (7,412)
CXBJ	2.921 $\pm$ 0.137 (883)
[(CXBJ) x B] F_1	3.280 $\pm$ 0.069 (1,906)
C	4.102 $\pm$ 0.078 (12,654)
CXBJ	2.921 $\pm$ 0.137 (883)
[C x (CXBJ)] F_1	4.118 $\pm$ 0.093 (13,132)

[a] Log_{10} PFC/spleen $\pm$ S.E.M. for 5-14 mice, 5 days after immunization with 0.5 μg SSS-III; geometric means are in parentheses.

The data of Table 8 summarize the results obtained which are expressed as the mean percent decrease in the magnitude of the PFC response to an optimally immunogenic dose (0.5 μg) of SSS-III after priming with a single injection of 0.005 μg of SSS-III 3 days before immunization. B and C mice did not differ significantly ($p > 0.05$) with respect to the degree of low-dose paralysis induced (80-84%); most strains of mice examined showed the same degree of unresponsiveness, designated by the symbol B/C for phenotype. This was true, regardless of H-2 type or IgC_H allotype, indicating that suppressor T cell activity is not linked to genes within either of these two genetic complexes. More important, the degree of low-dose paralysis induced in CXBI mice was uniquely low (49.1%), whereas that induced in CXBD mice was uniquely high (94.4%); the frequency of such unique phenotypes ($p < 0.05$ in both cases) suggests that at least 2 autosomal genes may govern the expression of suppressor T cell activity [21]. It should be noted that this may be a conservative estimate of the actual number of genes involved in the expression of suppressor T cell activity. Data with regard to the smallest dose of SSS-III needed to induce unequivocal unresponsiveness might have given different and more precise information along this line; but, sufficient numbers of each of the RI strains were not available to permit such an analysis to be made.

TABLE 8

DEGREE OF LOW-DOSE PARALYSIS INDUCED IN RI AND PROGENITOR B AND C MICE

	Degree of low-dose paralysis			
Mouse Strain	% decrease in PFC response [a]	Phenotype [b]	H-2 type	IgC_H allotype
CXBI	49.1 ± 12.3	Unique	B	B
CXBJ	62.7 ± 11.1	B/C	B	C
CXBG	70.0 ± 6.5	B/C	B	C
CXBE	72.7 ± 5.5	B/C	B	B
B	80.2 ± 7.6	B/C	B	B
C	84.2 ± 2.9	B/C	C	C
CXBK	87.3 ± 3.4	B/C	B	B
CXBH	90.2 ± 2.6	B/C	C	B
CXBD	94.4 ± 2.0	Unique	C	B

[a] Mean percent decrease in the PFC response ± S.E.M. for 9-13 mice to 0.5 μg SSS-III after pretreatment (priming) with 0.005 μg SSS-III. Mice were primed 3 days before immunization with 0.5 μg SSS-III.

[b] B/C indicates that the degree of low-dose paralysis found did not differ ($p > 0.05$) from that induced in either B or C mice. Unique responses were significantly different ($p < 0.05$) from those of both B and C mice.

The data of Table 9 show that the amount of enhancement produced after treatment with ALS, which provides a measure of the amount of amplifier T cell activity present, ranged from 6.5 - 190-fold. There was no direct relationship between the amount of enhancement found and either IgC_H allotype or H-2 type. With respect to H-2 type, the enhancement obtained with CXBH and CXBD mice was uniquely low; however, the enhancement noted for C mice having the same H-2 type was at least 10-15 times greater ($p < 0.05$). The enhancement for CXBJ and CXBI mice was uniquely high and more than 5-10 times greater ($p < 0.05$) than that of other strains with the same H-2 type. These unique responses, which occurred in 4 of the 7 RI strains examined, suggest that at least 3-4 autosomal genes may influence the expression of amplifier T cell activity.

In all of the data presented so far, the effects produced by suppressor and amplifier T cells were examined separately. It would be important to know if, on the basis of the results shown in Tables 7-9, one can make predictions concerning the ability of RI strains to respond to SSS-III. Do mice that have low suppressor T cell activity and high amplifier T cell activity give a high antibody response to SSS-III? The data of Table 10 review some of the more pertinent interactions observed in this regard. Both CXBJ and CXBI mice are _low_ in suppressor T cell activity and _high_ in amplifier T cell activity; one would, therefore, expect such a combination to result in the ability to give a high PFC response to SSS-III. Instead, both of these strains give low responses. The situation with CXBH and CXBD mice is almost the opposite; here, one would expect the combination of

TABLE 9

DEGREE OF ALS-INDUCED ENHANCEMENT OF THE PFC RESPONSE TO SSS-III IN RI AND PROGENITOR B AND C MICE

	Degree of enhancement			
Mouse Strain	Magnitude of Increase [a]	Phenotype [b]	H-2 type	IgC_H allotype
CXBH	6.5 ± 1.0	Unique	C	B
CXBD	8.7 ± 2.3	Unique	C	B
CXBG	13.4 ± 2.6	B	B	C
CXBK	14.9 ± 2.2	B	B	B
CXBE	21.1 ± 0.89	B	B	B
B	27.1 ± 0.89	B	B	B
C	82.1 ± 8.1	C	C	C
CXBJ	119.3 ± 14.4	Unique	B	C
CXBI	190.7 ± 6.8	Unique	B	B

[a] Mean increase in the PFC response ± S.E.M. for 9-12 mice given 0.3 ml ALS at the time of immunization with 0.5 μg SSS-III; PFC responses were assessed, 5 days after immunization and compared to those for non-ALS-treated mice.

[b] B and C indicate that the enhancement found was like ($p > 0.05$) that of B or C mice, respectively; unique signifies enhancement, significantly different ($p < 0.05$) from that of both B and C mice.

TABLE 10

INTERACTIONS BETWEEN GENES CONTROLLING THE EXPRESSION OF VARIOUS FUNCTIONAL ACTIVITIES [a]

Mouse strain	Suppressor T cell activity	Amplifier T cell activity	Capacity of B cells to respond to SSS-III
CXBJ	low	high	low
CXBI	low	high	low
CXBH	high	low	high
CXBD	high	low	intermediate

[a] The activities considered were classified on the basis of the data given in Tables 7, 8, and 9.

high suppressor T cell activity and low amplifier T cell activity to result in low responsiveness; but, these strains give either a high or an intermediate response. These observations suggest that the genes governing the expressions of amplifier and suppressor T cell activity interact in a complex -- and perhaps in a competitive -- manner, and that such genes may not necessarily act at the same level of control for the antibody response to SSS-III [21, 22]. It should be noted that there is no direct relationship between the ability of B cells to respond to SSS-III and the degree of suppressor and amplifier T cell activity present, as measured by the correlation coefficient obtained when values for these functions were ranked and compared statistically [22]. However, there is an inverse relationship between the amount of suppressor and amplifier T cell activity expressed;

thus, an increase in suppressor T cell activity is associated with a corresponding decrease in amplifier T cell activity [22]. This is not surprising in view of the fact that suppressor T cells are known to inhibit the activities of both amplifier T cells and B cells as discussed above and elsewhere [32].

It is clear, from the results of other studies [22], that the same RI strains of mice used in our work respond differently to other antigens that differ structurally from SSS-III; unique phenotypes are not restricted to any particular RI strain, nor are high (or low) antibody responses to one antigen invariably associated with high (or low) antibody responses to other unrelated antigens. It therefore appears likely that the genetic effects observed in our studies are characteristic (specific) for an antibody response to SSS-III and cannot be ascribed to general factors such as spleen size, absolute numbers of B cells, the rate at which immune functions may mature in a given RI strain, *etc*. At this time, it is not possible to define the gene products involved in the types of cellular interactions described in this work. To do so would require much more information than is now available on (a) the nature of the recognition units involved in the activation of regulatory T cells, as well as (b) the size and/or diversity of the T cell repertoire for the recognition of such structural units, be they idiotypic determinants or other types of receptor molecules. This represents one of the most challenging problems in immunology, one which is likely to occupy the attention of immunologists for quite some time. Our own particular point of view is that further genetic studies using other inbred strains of mice and different microbial antigens will permit one to identify many of the regulatory mechanisms involved and to relate the significance of such mechanisms to the prevention of disease and the survival of the species.

ACKNOWLEDGEMENTS

It has been a privilege to have had the opportunity to collaborate with several outstanding and creative young investigators during the course of all this work. All of them, whose work has been cited in the text and reviewed by us, have contributed significantly to any progress that has been made. They include: Dr. Norman D. Reed, Department of Microbiology, Montana State University, Bozeman, Montana; Dr. Richard B. Markham, Department of Medicine, Peter Bent Brigham Hospital, Boston, Massachusetts; and Dr. Jeffrey M. Jones, Department of Internal Medicine, Veterans Administration Hospital, Madison, Wisconsin. We are most grateful to Mr. Philip W. Stashak, Ms. Diana F. Amsbaugh, and Mr. George Caldes for their expert and professional assistance in completing the studies described.

REFERENCES

1. Humphrey, J.H., Parrot, D.M.V. and East, J. (1964) Immunology, 7, 419-436.

2. Davies, A.J.S., Carter, R.L., Leuchars, E., Wallis, V. and Dietrick, F.M. (1970) Immunology, 19, 945-957.

3. Howard, J.G. Christie, G.H., Courtenay, B.M., Leuchars, E. and Davies, A.J.S. (1971) Cell. Immunol. 2, 614-626.

4. Manning, J.K., Reed, N.D. and Jutila, J.W. (1972) J. Immunol. 108, 1470-1472.

5. Baker, P.J., Reed, N.D., Stashak, P.W., Amsbaugh, D.F. and Prescott, B. (1973) J. Exp. Med. 137, 1431-1441.

6. How, M.J., Brimacombe, J.S. and Stacey, M. (1964) Advance. Carbohy. Res. 19, 303-358.

7. Baker, P.J. (1975) Transplant. Rev. 26, 3-20.

8. Baker, P.J. and Stashak, P.W. (1969) J. Immunol. 103, 1342-1346.

9. Baker, P.J., Stashak, P.W., Amsbaugh, D.F. and Prescott, B. (1971) Immunology, 20, 481-492.

10. Baker, P.J., Stashak, P.W., Amsbaugh, D.F. and Prescott, B. (1971) Immunology, 20, 469-480.

11. Jones, J.M., Amsbaugh, D.F. and Prescott, B. (1976) J. Immunol. 116, 41-51.

12. Rittenberg, M.B. and Pratt, K.L. (1969) Proc. Soc. Exp. Biol. Med. 132, 575-581.

13. Jeanes, A., Wilham, C.A. and Miers, J.C. (1948) J. Biol. Chem. 176, 603-615.

14. Baker P.J., Stashak, P.W. and Prescott, B. (1969) Appl. Microbiol. 17, 422-426.

15. Leon, M.A., Young, N.M. and McIntire, K.R. (1970) Biochemistry 9, 1023-1030.

16. Baker, P.J., Stashak, P.W., Amsbaugh, D.F. and Prescott, B. (1974) J. Immunol. 112, 404-409.

17. Baker, P.J., Bernstein, M., Pasanen, V. and Landy, M. (1966) J. Immunol. 97, 767-777.

18. Baker, P.J. and Landy, M. (1967) J. Immunol 99, 687-694.

19. Markham, R.B., Reed, N.D., Stashak, P.W., Prescott, B., Amsbaugh, D.F. and Baker, P.J. (1977) J. Immunol. 119, 1163-1168.

20. Morse, H.C., III, Prescott, B., Cross, S.S., Stashak, P.W. and Baker, P.J. (1976) J. Immunol. 116, 279-287.

21. Baker, P.J., Amsbaugh, D.F., Prescott, B. and Stashak, P.W. (1976) J. Immunogen. 3, 275-286.

22. Baker, P.J., Amsbaugh, D.F., Prescott, B., Stashak, P.W. and Rudbach, J.A. (1978) in Infection, Immunity, and Genetics, Friedman, H., Prier, J. and Linna, T.J. eds., University Park Press, Baltimore, Md. pp. 67-83.

23. Markham, R.B., Stashak, P.W., Prescott, B., Amsbaugh, D.F. and Baker, P.J. (1978) J. Immunol. 121, 829-834.

24. Muirhead, D.Y. and Cudkowicz, G. (1978) J. Immunol. 121, 130-137.

25. Markham, R.B., Stashak, P.W., Prescott, B., Amsbaugh, D.F. and Baker, P.J. (1977) J. Immunol. 119, 1159-1162.

26. Jones, J.M., Amsbaugh, D.F., Stashak, P.W., Prescott, B., Baker, P.J. and Alling, D.W. (1976) J. Immunol. 116, 647-656.

27. Baker, P.J., Barth, R.F., Stashak, P.W. and Amsbaugh, D.F. (1970) J. Immunol 104, 1313-1315.

28. Baker, P.J., Stashak, P.W., Amsbaugh, D.F. and Prescott, B. (1974) J. Immunol. 112, 2020-2027.

29. Markham, R.B., Stashak, P.W., Prescott, B., Amsbaugh, D.F. and Baker, P.J. (1978) J. Immunol. 120, 986-990.

30. Baker, P.J., Burns, W.H., Prescott, B., Stashak, P.W. and Amsbaugh, D.F. (1974) in Immunological Tolerance: Mechanisms and Potential Therapeutic Applications, Katz, D.H. and Benacerraf, B., eds., pp. 493-502.

31. Rich, R.R. and Pierce, C.W. (1973) J. Exp. Med. 137, 205-223.

32. Rich, R.R. and Pierce, C.W. (1974) J. Immunol. 112, 1360-1368.

33. Markham, R.B., Stashak, P.W., Prescott, B., Amsbaugh, D.F. and Baker, P.J. (1977) J. Immunol. 118, 952-956.

34. Braley-Mullen, H. (1971) Cell. Immunol. 2, 73-81.

35. Gerety, R.J., Ferraresi, R.W. and Raffel, S.J. (1970) J. Exp. Med. 131, 189-206.

36. Lerman, S.P., Romano, T.J., Mond, J., Heidelberger, M. and Thorbecke, G.J. (1975) Cell. Immunol. 15, 321-335.

37. Baker, P.J., Prescott, B., Stashak, P.W. and Amsbaugh, D.F. (1971) J. Immunol. 107, 719-724.

38. Amsbaugh, D.F., Hansen, C.T., Prescott, B., Stashak, P.W., Barthold, D.R. and Baker, P.J. (1972) J. Exp. Med. 136, 931-949.

39. Amsbaugh, D.F., Hansen, C.T., Prescott, B., Stashak, P.W., Asofsky, R. and Baker, P.J. (1974) J. Exp. Med. 139, 1499-1511.

40. Bailey, D.W. (1971) Transplantation, 11, 325-327.

41. Swank, R.T. and Bailey, D.W. (1973) Science (Washington, D.C.), 181, 1249-1252.

42. Potter, M., Finlayson, J.S., Bailey, D.W., Muchinski, E.B., Reamer, B.L. and Walters, J.L. (1973) Genet. Res. 22, 325-328.

43. Potter, M., Pumphrey, J.G. and Bailey, D.W. (1975) J. Nat. Cancer Inst. 54, 1413-1417.

44. Braley, H.C. and Freeman, M.J. (1971) Cell. Immunol. 2, 73-81.

Rudbach/Baker, eds. Immunology of Bacterial Polysaccharides

REGULATORY EVENTS IN THE IMMUNE RESPONSE OF MICE TO DEXTRAN

MYRON A. LEON, JENN C. CHEN AND TUAN-HUEY KUO
Department of Immunology and Microbiology, Wayne State University School of Medicine, Detroit, Michigan, 48201

INTRODUCTION

Like humans [1], mice respond to immunization with dextrans. Antibodies with specificities to both α(1-6) and α(1-3) linked glucose are formed [2-6]. The number of responding PFC* are readily quantitated using sheep erythrocytes sensitized with O-acylated dextrans [7]. The PFC measured in response to immunization with dextrans are essentially all IgM. Production of significant amounts of 7S antibody requires combined immunization with dextran and *E. coli* B which share glucosyl determinants with the dextran [8].

This paper describes some aspects of the regulation of the antibody response to α(1-3) linked glucose determinants in dextrans. Many of the experiments utilize a T lymphocyte product, LDF, which drives partially differentiated PFC precursors to mature PFC in the absence of added antigen [9]. Overnight incubation of spleen cells from immunized animals, with or without LDF, provides a comparison of the numbers of precursors, relative to the number of mature PFC, at any time during the course of immunization.

---

*Abbreviations used in this paper: PFC, plaque forming cells; ^{3}H-TdR, tritiated thymidine; OSD, O-stearoyl dextran; OSDE, O-stearoyl dextran coated sheep erythrocytes; Con A, concanavalin A; LDF, lymphocyte differentiation factor; SIII, Type 3 pneumoccal polysaccharide.

MATERIALS AND METHODS

Dextran B1355 S was given by Dr. Allene Jeanes. OSD was prepared as previously described [7]. The medium for all cell cultures and cell washings was RPMI 1640 supplemented with 10% fetal bovine serum (Grand Island Biologicals, Grand Island, NY), penicillin (100 units/ml) and streptomycin (100 μg/ml). Female BALB/c mice (15 to 20 g) were supplied by Flow Research Animals, Inc. (Dublin, Va.). Nu/nu and nu/+ mice with BALB/c genetic backgrounds were gifts of Dr. R. K. Morrison. Con A was prepared by minor modifications of the method of Agrawal and Goldstein [10]. The conditions for *in vitro* stationary cell culture were described previously [6]. Cell viability was determined by trypan blue exclusion. Where indicated, hydroxyurea (Nutritional Biochemicals, Cleveland, Ohio) was added to cultures at a final concentration of 2 mM.

PFC were assayed by a modification of Jerne's hemolytic plaque assay [11]. Briefly, 1- to 2-week-old sterile sheep erythrocytes (Southwest Lab., Cleveland, Ohio) were washed four times with RPMI. Packed, washed sheep erythrocytes were sensitized with an equal volume of OSD (75 μg/ml) for 30 min at 37°C. Spleen cell suspension (0.1 ml), diluted to contain approximately 100 PFC, was mixed with 0.5 ml of 4% OSDE in 1.2% agarose which was prewarmed to 49°C. The slurry was poured on 1 x 3-inch slides precoated with 0.1% agarose which were then incubated for 1 hour at 37°C in 5% CO_2-95% air saturated with water vapor, then flooded with diluted (1:20) guinea pig complement (Grand Island Biological Co.). PFC were enumerated after 2 hours of additional incubation. The background PFC against sheep erythrocytes were subtracted from experimental PFC against OSDE.

Indirect PFC were assayed by a previously described method [11], using goat anti-mouse IgG (Meloy Laboratories, Inc., Springfield, Va.) which was dialyzed against RPMI and absorbed with sheep erythrocytes before use. The slides were incubated for 1 hour and then flooded with diluted anti-IgG serum for 1 additional hour. The slides were drained and treated with guinea pig complement as above and reincubated for 2 more hours. The data are expressed as PFC/10^6 splenic lymphocytes in all experiments where no *in vitro* culture was performed.

Assays for incorporation of ^{3}H-TdR were performed by adding ^{3}H-TdR (Amersham-Searle, Chicago, Il.) to cultures at a final concentration of 0.7 μCi per ml. After a four hour pulse, 0.1 ml aliquots of cultures were harvested with a cell harvester (Otto Hiller Co., Madison, Wi.). The method for preparation and counting of samples has been previously described [12].

LDF was prepared by stationary culturing of either splenic lymphocytes or thymocytes (5 x 10^6/ml) for 40 hours (37°C, 95% air-5% CO_2) in RPMI 1640 without fetal calf serum. At the inception of culture, 1.6 μg of Con A were added/ml of culture medium. Cells were centrifuged after 36-40 hours and the supernatant passed through Sephadex G-200 (Pharmacia, Piscataway, NJ). The excluded fraction, containing the LDF activity, was sterilized by filtration and stored in small aliquots at -70°C. Methyl-α-D-mannoside (0.04M) was routinely added to all test systems using LDF to inhibit traces of aggregated Con A which might contaminate the LDF preparations. LDF (0.1 ml) was added to cultures containing 1 x 10^7 spleen cells in a volume of 2 ml. Cultures were incubated for the indicated times at 37°C in 95% air-5% CO_2. Cells were harvested by centrifugation, washed twice and assayed for PFC as described above. Data for all _in vitro_ cultures involving LDF are expressed as PFC/10^7 cells.

RESULTS

The _in vivo_ immune response of conventional BALB/c mice to dextran was studied after injecting dextran B1355 S intravenously. The number of splenic anti-dextran PFC rises (Fig. 1) to a maximum, at all doses tested, 5 days after immunization. At the maximal response, obtained with a dose of 20 μg dextran, approximately 0.05% of the total splenic leukocyte population is measurable as anti-dextran PFC. A dose of 0.8 μg stimulates a smaller response than that observed with either 20 μg or 500 μg of dextran. Although there is a 600-fold difference between the lowest and highest priming doses, no apparent suppression of PFC responses is observed at the highest dose used in these experiments. Doses higher than 1 mg per mouse kill significant numbers of the mice.

After five days the number of PFC declines rapidly. However, a persistent low level of PFC remains and, even after eight months, 20 to 40 PFC per million spleen cells are still detectable. No indirect plaques are detected at any time during the response. These results are similar to those obtained using the structurally closely related parent dextran B1355 [4], although higher doses of the latter (1 mg) were required for optimal response.

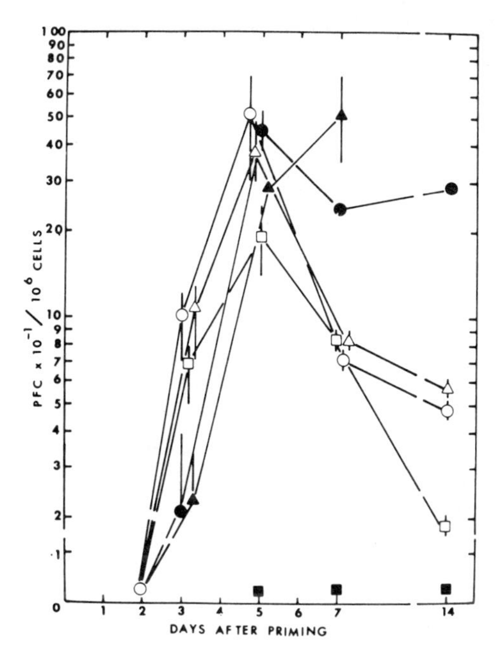

Fig. 1. The *in vivo* PFC response of conventional BALB/c and nu/nu-BALB/c to immunization with dextran B1355S i.v. on day 0. Mice were sacrificed on the days indicated. Conventional mice immunized with 500 μg, △ ; 20 μg, ○; 0.8 μg, □ ; nu/nu mice immunized with 500 μg, ▲ ; 20 μg, ●; non-primed controls and indirect plaques, ■ . Vertical bars indicate standard deviations. Standard deviations are not shown for nu/nu experimental points where only 2 animals were used. (Reprinted from reference 6 with permission of Williams and Wilkins Co.).

Mice, rechallenged with dextran, fail to respond normally. The magnitude of response to the rechallenge increases as the interval between injections increases. However, full recovery does not occur even after a 15 week interval (Table 1).

TABLE 1

PERSISTENCE OF REFRACTORY STATE TO RECHALLENGE WITH DEXTRAN

First dose	Second dose	PFC[a]
day	day	
0	112	244 ± 11
70	112	87 ± 5
105	112	21 ± 6
112	-	599 ± 25

[a] PFC measured on day 117. All injections were 20 μg of dextran i.v.

Since Baker et al.[12,13] have demonstrated a role for T suppressor cells in limiting the response of mice to SIII, it was logical to suspect that suppressor T cells might be involved in regulating the anti-dextran response, viz., magnitude, duration, and refractory state to rechallenge. If this hypothesis were true, the response of nu/nu mice, lacking significant numbers of mature T cells, should differ markedly from that of conventional mice. In fact, response of nu/nu mice to intravenous injection of 20 μg of dextran reaches a maximum at 5 days, and then, unlike the response of conventional mice persists at a high level for the duration of the experiment (14 days) (Fig. 1). When nu/nu mice are reconstituted with 2.5×10^7 T cell-enriched spleen cells from nu/+ littermates, just prior to injection of dextran, their response parallels that of conventional mice, i.e., the number of PFC declines rapidly after 5 days (Table 2).

TABLE 2

RESPONSE TO DEXTRAN OF NU/NU MICE RECONSTITUTED WITH NU/+ T CELLS

Mice	PFC[a]
nu/nu	527 ± 62
reconstituted nu/nu[b]	169 ± 45
nu/+	210 ± 90

[a] PFC were assayed 8 days after dextran injection.

[b] Spleen cells from nu/+ littermates were passed through a nylon fiber column and injected (2.5 x 10^7, i.v.) into nu/nu recipients prior to injection of 20 μg of dextran.

These data demonstrate that although the response to dextran is "T independent" in the sense that few, if any, T cells are required for production of IgM anti-dextran, T suppressor cells regulate the amplitude and duration of the response. Furthermore, since nu/nu mice maintain their high level of PFC in the absence of T cell help, the proliferation and differentiation processes induced by dextran do not appear particularly sensitive to feedback inhibition by the IgM antibody being produced during this period.

Data to be published elsewhere demonstrate that the reconstitution of the nu/nu mice must be accomplished within 48 hours after the injection of dextran for their response to parallel that of conventional mice. The number of PFC recovered 7 days after injection of dextran decreases as the interval between reconstition and injection of dextran increases. Thus, the generation of suppressor cells in this system must take place relatively soon after immunization. Whether IgM PFC precursors can only be suppressed relatively early in their differentiation or whether suppressor cells can only be generated at a particular stage of antigen processing is not yet known.

Comparison of the kinetics of the response of nu/nu and conventional mice shows that at day 3, the number of PFC measured in the nu/nu mice is significantly lower than that measured in the conventional animals. These data suggested the possibility that, in addition to their suppressive effects, T cells can also provide help for the "T independent" response to dextran. In the initial experiments, spleen cells obtained from normal or nu/nu mice, primed with either 500 μg or 20 μg of dextran 3 days earlier, were cultured for 2 days in vitro with or without Con A (1.1 μg/ml culture). Table 3 shows that Con A has no effect on the PFC response of the nu/nu mice but substantially enhances the response of conventional mice.

The factor involved in this enhancement, LDF, was subsequently prepared from either thymocytes or spleens of conventional unimmunized animals as described in Materials and Methods. The fact that LDF is produced by thymocytes or spleen cells of conventional mice and not by spleen cells from nu/nu mice indicates the T cell origin of the factor.

Although not produced by spleen cells from nu/nu mice, LDF drives the differentiation of precursors of anti-dextran PFC taken from nu/nu or conventional mice [9]. Thus, when spleen cells from dextran-primed mice are removed at day 5 and cultured overnight in the presence of LDF, the number of anti-dextran PFC recovered is markedly increased (Table 4).

TABLE 3

EFFECT OF CON A ON ANTI-DEXTRAN RESPONSE OF PRIMED CELLS FROM NU/NU OR CONVENTIONAL MICE.

Cell source	Priming dose[a]	Additions[b]	PFC/culture[c]
	µg		
Athymic	20	None	322 ± 19
Athymic	20	Con A	306 ± 14
Athymic	500	None	177 ± 49
Athymic	500	Con A	165 ± 31
Normal	20	None	1600 ± 130
Normal	20	Con A	3320 ± 260
Normal	500	None	298 ± 20
Normal	500	Con A	857 ± 38

[a] Mice were primed with dextran for 3 days.

[b] Con A (1.1 µg/ml) was added at initiation of culture.

[c] Cultures were set up with 2 ml of pooled spleen cells (5 x 10^6 cells/ml). Each number represents mean ± S.D. from triplicate cultures. PFC were assayed 2 days after Con A addition.

TABLE 4

LDF ENHANCES DIFFERENTIATION OF PRECURSOR CELLS TO MATURE ANTI-DEXTRAN PFC[a]

LDF	PFC
+	3047 ± 150
-	395 ± 24

[a] Spleen cells from mice primed 5 days earlier with 20 μg dextran were washed and cultured 18 hours with LDF before assay. No dextran was added to the *in vitro* culture.

We have used hydroxyurea to determine whether the increase in PFC observed on incubation of primed cells with LDF is due to differentiation alone or differentiation in conjunction with clonal proliferation. Hydroxyurea arrests passage of cells from G1 to S and; in addition, it kills cells already actively engaged in DNA synthesis [15]. Spleen cells from dextran-primed mice were removed at 5 days and cultured overnight with LDF in the presence or absence of hydroxyurea. Table 5 shows that there is essentially no difference in the enhancement of PFC numbers by LDF whether hydroxyurea is, or is not added. Moreover, the experiment demonstrates that the proliferative phase for cells taken from animals primed for 5 days is essentially finished since hydroxyurea does not substantially diminish the PFC yield from the cultures lacking LDF. However, the same concentration of hydroxyurea completely blocks thymidine incorporation into cells stimulated by Con A (Table 6), i.e., it blocks processes involving DNA synthesis and proliferation. The data support the interpretation that LDF drives differentiation of precursors to mature PFC in the absence of cell division.

TABLE 5

EFFECT OF HYDROXYUREA ON DIFFERENTIATION OF PFC INDUCED BY LDF[a]

LDF	OH urea	PFC
-	-	395 ± 24
-	+	357 ± 20
+	-	3047 ± 150
+	+	2880 ± 72

[a] 1×10^7 spleen cells from mice primed i.v. 5 days previously with 20 μg B1355S were incubated for 20 hours with reagents indicated, *in vitro*, prior to assay for PFC.

TABLE 6

EFFECT OF HYDROXYUREA ON THYMIDINE INCORPORATION[a]

CON A	OH urea	CPM
+	-	141,320 ± 23,761
+	+	308 ± 89
-	-	230 ± 52

[a] 2.5×10^5 spleen cells incubated with Con A (1 μg/ml) for 2 days when hydroxyurea was added for 18 hours followed by pulsing (4 hours) with ^{3}H-TdR and harvesting.

The time required for the process of differentiation of precursors to mature PFC to become independent of added LDF was investigated as follows. Five day primed spleen cells were cultured with LDF for various times, centrifuged, washed, resuspended in medium without LDF and placed in culture again. After 20 hours total culture time (with and without LDF), the cells were harvested and the number of PFC measured. The results, presented in Table 7, show that LDF must be present throughout the culture period to obtain the maximal yield of mature PFC.

TABLE 7

PRECURSORS REQUIRE THE CONTINUOUS PRESENCE OF LDF FOR DIFFERENTIATION TO MATURE PFC[a]

Time incubated with LDF	PFC
hr	
0	70 ± 30
0.25	140 ± 15
1	300 ± 25
5	625 ± 30
20	1005 ± 65

[a] 1×10^7 spleen cells from mice primed with 20 μg B1355S 5 days, incubated with LDF, washed twice and resuspended in medium without LDF for a total incubation time of 20 hours.

DISCUSSION

From the data presented in this paper and earlier work [6], the events occurring during the immune response to dextran of conventional BALB/c mice can be divided into three stages, early ($<$ 5 days), middle (5 days) and late ($>$ 5 days). During the early stage, the number of terminally differentiated cells (PFC) is rapidly increasing. This increase comes from the extensive proliferation and differentiation of precursor cells. Evidence for the domination of proliferative events comes from experiments showing that hydroxyurea blocks expression of the majority of PFC in short term cultures of spleen cells taken from animals during this period [6]. By day 5, proliferation of precursors has essentially finished (Table 5) and the number of PFC measured in the spleen reaches an asymptote. Since the lifetime of a murine plasma cell is approximately 12 hours [16], in the absence of significant proliferation, only a high rate of differentiation of precursors could maintain the level of PFC. The fact that the number of splenic PFC falls rapidly after 5 days indicates that, during the late phase of the response, sufficient numbers of precursors do not differentiate to PFC.

That there are large numbers of precursors that do not normally differentiate to mature PFC in 5 day primed spleen is demonstrated by the experiments with LDF. Depending on the preparation of LDF used, from three to ten-fold enhancement of PFC is induced after 18 hours of *in vitro* culture. Since hydroxyurea does not block, in this case, the enhancement must involve differentiation without cell proliferation, and the number of precursor cells must at least equal the increase in number of PFC.

Why do these precursors not differentiate to mature PFC? They may be blocked by the direct action of suppressor T cells, blocked indirectly by the lack of sufficient T help, or, antigen differentiation may be an intrinsically inefficient process leaving large numbers of cells in various compartments along the differentiation pathway.

Nu/nu mice, which maintain a high level of PFC during the period when PFC in conventional mice have fallen to low levels, are converted to conventional behavior by reconstitution with T cells. These observations demonstrate that T suppressor cells act to limit the anti-dextran response of conventional mice. These observations, however, do not provide evidence for the level or levels at which the suppression is manifested. Thus, suppressor activity could affect proliferative and/or differentiative processes. Projected studies of the effects of hydroxyurea on short-term cultures of nu/nu spleen cells primed with dextran for different times should provide more definitive data.

While it is apparent that T cell help (LDF) can drive precursors to mature PFC in the experimental situation of *in vitro* culture, there is no direct evidence that T cell help can be induced by dextran *in vivo*. There are, however, two observations that may be used as indirect evidence that T cell help does play a role in the anti-dextran response of conventional animals. First, the fact that the rate of appearance of anti-dextran PFC in the nu/nu mouse lags behind that of conventional mice (Fig. 1). Second, the fact that the maximal response of conventional mice is virtually identical to that of nu/nu mice despite the presence of active suppressor cells. The latter argument played an important role in the development of a model for response to SIII [13,14,17].

Finally, the ability of LDF to detect precursor cells may be used to examine the numbers of such cells in conventional and nu/nu mice during responses to a variety of antigens. Such studies may indicate that inefficiency of antigen-driven processes provides important biological advantages such as memory cell production.

ACKNOWLEDGMENT

This work was supported in part by Grant CA 20812 from the National Cancer Institute, Department of Health, Education and Welfare.

REFERENCES

1. Kabat, E.A. and Berg, D. (1953) J. Immunol. 70, 514-532.

2. Blomberg, B., Geckeler, W.R. and Weigert, M. (1972) Science, 177, 178-180.

3. Howard, J.G., Vicari, G. and Courtenay, B.M. (1975) Immunoolgy, 25, 585-597.

4. Howard, J.G. and Courtenay, B.M. (1975) Immunology, 25, 599-610.

5. Howard, J.G., Courtenay, B.M. and Vicari, G. (1975) Immunology, 25, 611-619.

6. Chen, J.C. and Leon, M.A. (1976) J. Immunol. 116, 416-422.

7. Leon, M.A., Young, N.M. and McIntire, K.R. (1970) Biochemistry, 9, 1023-1030.

8. Hansburg, D., Briles, D.E. and Davie, J.M. (1976) J. Immunol. 177, 569-575.

9. Kuo, T.-H., Tartamella, R. and Leon, M.A. (1978) Fed. Proc. 37, 1572.

10. Agrawal, B.B.L. and Goldstein, I.J. (1965) Biochem. J. 96, 23c.

11. Shearer, G.M., Cudkowicz, G., Connell, S.J. and Priore, R.L. (1968) J. Exp. Med. 128, 437-457.

12. Powell, A.E. and Leon, M.A. (1970) Exp. Cell. Res. 62, 315-325.

13. Baker, P.J., Stashak, P.W., Amsbaugh, D.F., Prescott, B. and Barth, R. (1970) J. Immunol. 105, 1581-1583.

14. Baker, P.J., Reed, N.D., Stashak, P.W., Amsbaugh, D.F. and Prescott, B. (1973) J. Exp. Med. 137, 1431-1441.

15. Sinclair, W.K. (1965) Science, 150, 1729-1731.

16. Schooley, J.C. (1961) J. Immunol. 86, 331-337.

17. Markham, R.B., Reed, N.D., Stashak, P.W., Prescott, B., Amsbaugh, D.F. and Baker, P.J. (1977) J. Immunol. 119, 1163-1168.

Rudbach/Baker, eds. Immunology of Bacterial Polysaccharides

IMMUNOGENICITY OF MENINGOCOCCAL POLYSACCHARIDES IN MAN

RONALD GOLD
Department of Pediatrics, University of Connecticut School of Medicine, Farmington, Connecticut 06032

INTRODUCTION

Safe and effective vaccines consisting of purified capsular polysaccharides have been prepared from Groups A and C meningococci. The Group A vaccine has been shown to protect infants [1], children [2-4], and adults [5] while the Group C vaccine prevents disease in children over 2 years of age [6] and adults [7-9]. The purpose of this report is to review the safety and immunogenicity of the meningococcal polysaccharide vaccines in man. Most of the data to be presented originate from our studies conducted in Danbury, Ct. over the past 7 years [10-17]. The infants and children who participated in these evaluations of Groups A and C meningococcal polysaccharide vaccines were recruited by asking parents of newborns and of school children to enroll in the various trials. None of these studies would have been possible without the outstanding cooperation of the participants and their parents, as well as the physicians, the Department of Health, the Danbury Hospital, and the Danbury school system.

CHEMISTRY OF MENINGOCOCCAL CAPSULAR POLYSACCHARIDES

Meningococci can be differentiated into 9 serogroups by means of agglutination reactions. The capsular polysaccharides are the antigens responsible for serogroup specificity; their chemical compositions are indicated in Table 1. Group C meningococci have

TABLE 1

CHEMISTRY OF MENINGOCOCCAL CAPSULAR POLYSACCHARIDES [a]

Serogroup	Repeating Unit of Polysaccharide
A	N-acetyl, O-acetyl mannosamine-6-phosphate (α-1-6)
B	N-acetyl neuraminic acid (α-2-8)
C	N-acetyl, O-acetyl neuraminic acid (α-2-9)
C variant	N-acetyl-neuraminic acid (α-2-9)
D	Unknown
X	N-acetyl glucosamine-4-phosphate (α-1-4)
Y	N-acetyl, O-acetyl neuraminic acid: glucose (α-2-6)
Z	Unknown
W135	N-acetyl neuraminic acid: glucose (α-2-6)
29E	3-deoxy-octulosonic acid: 2-deoxy-galactosamine

[a] See reference 19 for complete details.

been found to have two slightly different capsular polysaccharides[18]. Most Group C strains have both N- and O-acetyl groups in the polysaccharide which is resistant to enzymatic degradation by neuraminidase. The C-variant polysaccharide found on a few group C strains lacks O-acetyl groups and is sensitive to neuraminidase[19]. Vaccines have been prepared from Groups A, B, C, and Y meningococci and evaluated for safety and immunogenicity in man.

PREPARATION OF MENINGOCOCCAL POLYSACCHARIDES VACCINES

Gotschlich, *et al*. have developed a relatively simple

procedure permitting the purification of large molecular weight polysaccharides[20]. Following aerobic growth in fluid medium, the meningococci are killed and the polyanionic polysaccharides are precipitated by the addition of the cationic detergent Cetavlon (cetrimonium bromide). The precipitate is resuspended in water and the polysaccharide dissolved in calcium chloride. Nucleic acids and bacterial debris are removed by the addition of 25% ethanol followed by centrifugation. The polysaccharide is then precipitated from the supernate by increasing the ethanol concentration to 80%. Residual protein is removed by cold-phenol extraction of group A and by chloroform-butanol for group C; endotoxin is removed by ultracentrifugation.

TABLE 2

SPECIFICATIONS FOR GROUP A AND C MENINGOCOCCAL POLYSACCHARIDE VACCINES [a]

Analysis	Specification
Chemical	A: mannosamine phosphate $> 60\%$ or inorganic phosphorous $> 7.5\%$ C: sialic acid $> 80\%$ O-acetyl content $> 2 \mu mol/mg$ Protein $< 1\%$; Nucleic acid $< 1\%$
Physical	At least 50% of the polysaccharide recovered from Sepharose 4B column must have exclusion coefficient (Kd) of 0.5 or smaller.
Serological	Purity: no inhibition by vaccine of hemagglutination of erythrocytes sensitized with heterologous antigen in presence of heterologous antibody. Specificity: inhibits hemagglutination of erythrocytes sensitized with homologous antigen in presence of homologous antibody.
Biological	Rabbit pyrogenicity assay for endotoxin.

[a] See reference 21 for complete details.

The standardization of the final product depends on a variety of physico-chemical, serological, and biological assays. Because there is no animal model in which to test immunogenicity, the potency of each lot of vaccine is determined according to the specifications listed in Table 2 which have been shown to indicate safety and immunogenicity in man. These specifications have been developed in order to avoid having to test every lot in human volunteers[21-23].

DETECTION OF ANTIBODIES TO MENINGOCOCCAL POLYSACCHARIDES

Antibodies to the capsular polysaccharides as well as other surface antigens such as the outer membrane proteins and lipopolysaccharides can be measured by a variety of serological techniques[24]. Bactericidal, indirect hemagglutination, indirect-immunofluorescence and radioactive antigen binding assays have been used most extensively. The latter has proven to be the most sensitive method. The assay, as performed by Dr. E. C. Gotschlich, provides results in terms of μg of antibody protein per ml of sera[25]. The results to be described are expressed in terms of geometric mean antibody levels because, in the range of 10 to 90% antigen binding, the % antigen bound is linearly related to the logarithm of the antibody concentration.

NATURAL IMMUNITY TO MENINGOCOCCAL CAPSULAR POLYSACCHARIDES

Natural resistance to invasive disease caused by meningococci has been shown to correlate with the presence of bactericidal antibody[26]. The pattern of acquisition of antibody to the groups A and C meningococcal polysaccharides is shown in Table 3. Most

TABLE 3

NATURAL IMMUNITY TO GROUPS A AND C MENINGOCOCCAL POLYSACCHARIDES

Age	Anti-A Concentration (μg/ml)			Anti-C Concentration (μg/ml)		
	N	G.M.[a]	% > 0.13[b]	N	G.M.[a]	> % 0.11[b]
2mo	96	0.48	77	96	0.13	11
3mo	234	0.32	62	236	0.11	9
4mo	78	0.27	55	239	0.11	9
6mo	15	0.28	73	124	0.11	11
7mo	83	0.18	64	151	0.12	13
12mo	83	0.24	69	202	0.15	25
18mo	45	0.69	95	46	0.15	48
2yr	21	0.76	100	50	0.16	46
3yr	22	1.00	100	17	0.16	47
4yr	20	1.72	100	19	0.23	79
5yr	20	2.23	100	19	0.39	89
6-8yr	17	2.50	100	20	0.34	100
9-11yr	13	3.77	100	18	0.55	94
12-14yr	10	6.13	100	14	1.61	100

[a] Geometric mean antibody concentration, μg of antipolysaccharide antibody per ml of serum for N individuals, determined by radioactive antigen binding assay.

[b] % with detectable antibody: > 0.13 μg/ml of anti-A and > 0.11 μg/ml of anti-C.

infants have detectable anti-A antibody at 3 months of age which was acquired transplacentally from their mothers. In contrast, very few infants have detectable anti-C antibody at even 2 months of age. The subsequent rates of acquisition of anti-A and anti-C antibody are significantly different. Anti-A is acquired earlier and in much greater concentration than anti-C antibody. Indeed, anti-C concentrations are very low until 5-6 years of age whereas the geometric mean anti-A concentration reaches 1-2 μg/ml by 18-24 months of age. Similar differences are seen in the proportion of children at different ages who have detectable anti-A and anti-C antibodies. The proportion of children with detectable anti-A increases rapidly after 7 months of age and exceeds 90% by 18 months of age. Not until 6-8 years of age does a similar proportion of children have $\geq$0.11 μg/ml of anti-C antibody.

The sources of the antigens responsible for this natural immunization to groups A and C meningococcal polysaccharides have not been identified. Clearly it is unlikely that the group A and group C meningococci are involved directly since they are rarely found in healthy infants and acquisition of virulent meningococci by individuals without antibody usually results in disease. Nasopharyngeal carriage of meningococci was infrequent in infancy and early childhood[27]. The acquisition rate of meningococci during the first year of life was 2.5% and was 3.9% per year in children 6-8 years of age. The majority of the meningococcal strains were non-groupable or rough; no group A or C meningococci were found among 557 infants cultured 6 times during the first year of life or among 1084 6-8 year old children cultured 5 times over a 1 year period. Carriage of groupable meningococci did not become common until adolescence.

Carriage of bacteria in the throat and intestinal tract which have antigens cross-reactive with groups A, B, and C meningococci as well as *H. influenzae* type b has been demonstrated[28]. The role of such crossreacting bacteria in the development of natural immunity to meningococci and other pyogenic bacteria is under active investigation. As will be discussed subsequently, the immunogenicity of the purified group A meningococcal polysaccharide correlates with preimmunization antibody levels, i.e., with the prior experience of the host's immune system with these or similar antigens. Thus the rate of development of natural immunity may be an important determinant of the antibody responses to the vaccines observed at different ages.

SAFETY OF MENINGOCOCCAL POLYSACCHARIDE VACCINES

The safety of both the A and C vaccines has been clearly established during extensive field trials in the United States, Brazil, the Sudan, Egypt, and Finland[1-9]. Significant local or systemic reactions have not been observed in infants, children, or adults except with a few lots of A vaccine used in Finland (see below). Transient local erythema at the site of injection and irritability were observed in 8-10% of American infants receiving the A or C vaccines (Table 4)[12,17]. Reactions are less frequent in older children and adults. Local reactions are more common after intracutaneous administration than after subcutaneous injection by either needle and syringe or jet injector[29].

TABLE 4

ADVERSE REACTIONS OBSERVED IN INFANTS IMMUNIZED WITH GROUP A AND C MENINGOCOCCAL POLYSACCHARIDE VACCINES

		Type of Reactions (%)		
Age(mo)	Number	None	Local[a]	Systemic[b]
3	495	91	2	7
7	327	88	4	8
12	367	95	2	3
24	316	96	1	3

[a] Local = erythema of less than 2.5 cm lasting less than 24 hours.

[b] Systemic = irritability or lethargy lasting less than 24 hours.

During the recently concluded field trial of the A vaccine in Finland, 1.8% of infants had fever $>$ 38.5°C[1]. Subsequent analysis of the lots of A vaccines showed a significant correlation between the endotoxin concentration and the reaction rates[30]. The association between febrile reactions and endotoxin content indicates the need for strict quality control in the production of the vaccines.

A polysaccharide vaccine against group Y meningococci has been evaluated in 90 adults[31]. No systemic reactions occurred. Transient mild local reactions were reported.

IMMUNOGENICITY OF MENINGOCOCCAL POLYSACCHARIDE VACCINES

Group A Meningococcal Polysaccharide

The immunogenicity of the groups A and C meningococcal polysaccharide vaccines has been shown to depend on a number of variables. Host factors include the age and prior experience with the antigens as a result of natural exposure or immunization. Important vaccine variables include the dose of polysaccharide, molecular size of vaccine, and perhaps, the interval between immunizations. The route of administration does not appear to affect the antibody response. No significant differences in antibody levels were observed after immunization by the intracutaneous or subcutaneous routes with needle and syringe or by the jet injector gun[29]. Because the responses to the groups A and C polysaccharides are significantly different, the two vaccines will be discussed separately.

TABLE 5

ANTIBODY RESPONSES OF INFANTS AND CHILDREN AFTER PRIMARY IMMUNIZATION WITH GROUP A MENINGOCOCCAL POLYSACCHARIDE VACCINE

	Anti-A Concentration (μg/ml)				
	Pre[a]		Post[b]		
Age	N	G.M.[c]	N	G.M.	%>2.0 μg/ml[d]
3mo	99	0.34	119	0.30	3
7mo	71	0.18	76	0.39	4
12mo	55	0.25	56	1.22	36
18mo	61	0.65	65	3.64	77
2-5yr	34	0.78	33	5.25	88
6-8yr	91	1.68	127	9.35	98

a Pre = Pre-immunization level of serum antibody.

b Post = Antibody level 1 month after immunization.

c G.M. = Geometric mean anti-A concentration (μg/ml of serum) for N individuals, determined by radioimmunoassay.

d Percent of children with > 2.0 μg/ml of anti-A antibody 1 month after immunization.

Age. The response to primary immunization with the Group A polysaccharide at different ages is indicated in Table 5[16,17]. No antibody response is detected at 3 months of age presumably because of the presence of significant amounts of transplacental anti-A antibody. Thereafter, with increasing age, there is a significant increase in the peak antibody concentration achieved one month post-immunization. Indeed, between 7 months and 21 years of age, there is a linear relationship between peak anti-A concentration and the logarithm of age. By 18 months of age, primary immunization induces geometric mean anti-A concentration of 3.04 μg/ml; 77% of children have $\geq$2.0 μg/ml of antibody. Although the minimum antibody concentration required for protection against disease has not been determined, the protective level has been estimated to be approximately 2.0 μg/ml. Sixty percent of adults have $\geq$2.0 μg/ml of anti-A antibody and also have bactericidal activity against the group A meningococcus[1]. Moreover, sera obtained from patients at the time of admission with group A meningococcal disease had anti-A levels less than 2.0 μg/ml[5]. Finally, protection against disease was observed in immunized infants in whom a geometric mean anti-A concentration of 2 μg/ml or more was induced[1].

Prior Experience with Group A Polysaccharide. The increase in response to primary immunization with age may relate in part to the increase in prevalence of naturally acquired immunity to the group A polysaccharide with age. At 7 and 12 months of age, the geometric mean anti-A concentration after immunization was significantly greater in infants who had detectable anti-A antibody before receiving the vaccine than in infants without detectable antibody (Table 6)[12].

TABLE 6

CORRELATION BETWEEN PRE-IMMUNIZATION CONCENTRATION OF ANTIBODY TO GROUP A POLYSACCHARIDE AND RESPONSE TO THE GROUP A MENINGOCOCCAL POLYSACCHARIDE VACCINE

Age(mo)	Anti-A Concentration (μg/ml)[a]	
Pre-immunization	$<$ 0.13	$\geq$0.13
7	0.24	0.64
12	0.30	1.63

[a] Geometric mean anti-A concentration 1 month post-immunization.

TABLE 7

ANTIBODY RESPONSES OF INFANTS OF DIFFERENT AGES TO PRIMARY AND BOOSTER INJECTIONS OF GROUP A MENINGOCOCCAL POLYSACCHARIDE VACCINE

Age	Immunization[a]	Anti-A Concentration (μg/ml)[b]	%$>$2.0 μg/ml[c]
7mo	1°	0.39	4
	2°	2.51	61
12mo	1°	1.22	36
	2°	4.00	79
2-6yr	1°	13.23	100
	2°	14.16	100

[a] Primary (1°) immunization took place at the age given in the first column. For booster (2°) immunization, infants receiving a booster injection at 7 and 12 months of age had primary immunization at 3 months of age; children 2-6 years of age received a booster 3 months after primary immunization.

[b] The geometric mean anti-A antibody concentration.

[c] Percent of children with $>$ 2.0 μg/ml of anti-A antibody 1 month after immunization.

Just as natural priming of the immune system enhanced the response to the group A polysaccharide, true anamnestic responses to booster doses of vaccine were seen in infants (Table 7). When infants who had received primary immunization at 3 months of age were given boosters of A vaccine at 7 or 12 months of age, significantly greater antibody levels were induced than after primary immunization at those ages[12]. The geometric mean anti-A concentrations after primary or booster immunization at 7 months of age were 0.39 and 2.51 μg/ml respectively and at 12 months of age were 1.22 and 4.00 μg/ml respectively. Similar antibody responses were observed in Finnish infants given 2 doses of A vaccine 3 months apart[1]. Thus, even though no antibody response is detected after the primary immunization at 3 months of age, the immune system of the infant is altered so that a significant anamnestic response occurs to a subsequent dose of A vaccine.

The importance of the anamnestic response of young infants to 2 doses of group A vaccine is seen when the proportion of infants having ≥2.0 μg/ml of anti-A antibody is examined. Following primary immunization at 7 months of age, 4% of infants have ≥2.0 μg/ml compared to 61% who had received vaccine at 3 and 7 months of age. At 12 months of age, 36% of infants receiving primary immunization achieve ≥2.0 μg/ml of anti-A compared to 79% after vaccine at 3 and 12 months of age.

By 2 years of age, anamnestic responses are no longer observed. Children given a second or third dose of A vaccine following prior immunization with 1 or 2 doses of vaccine between 2 and 12 months of age have post-immunization anti-A levels no different than children receiving primary immunization[14,17]. The antibody concentrations induced in children 2-10 years of age by 2 doses of A vaccine administered 2 months apart were no greater than those seen after a single dose of vaccine[17]. The lack of booster responses in older children probably reflects the fact that natural immunization has already primed their immune systems so that a maximal response is obtained with the first immunization.

Dose of Group A Polysaccharide. The responses of infants to different doses of group A meningococcal polysaccharide is indicated in Table 8. No significant differences in the antibody responses to 25, 50, or 100 µg of vaccine were seen following primary immunization at 7, 12, or 18 months of age[12,17]. This may be of importance in terms of availability of vaccine during epidemics since twice as many doses of 25 µg are available from the same quantity of vaccine as with the standard 50 µg dose.

Molecular Size of Group A Polysaccharide. A major determinant of the immunogenicity of the meningococcal polysaccharides is the molecular size which is determined by gel filtration on Sepharose 4B columns[21,23]. Because the molecular size distribution of the polysaccharides is polydispersed, a single molecular weight is not identifiable. Molecular size of the polysaccharides is therefore expressed in terms of the partition coefficient, Kd, which is inversely related to molecular size: the smaller the value of Kd, the larger the average molecular size. Kd is calculated from the equation:

$$Kd = \frac{Ve - Vo}{Vi}$$

where Ve is elution volume measured to the peak of the polysaccharide elution curve, Vo is the void volume determined with Blue

TABLE 8

ANTIBODY RESPONSES OF INFANTS OF DIFFERENT AGES TO PRIMARY IMMUNIZATION WITH DIFFERENT DOSES OF GROUP A MENINGOCOCCAL POLYSACCHARIDE VACCINE

Age(mo)	Vaccine Dose (μg)	Anti-A Concentration (μg/ml) [a]
7	25	0.37
	100	0.40
12	25	0.77
	100	1.07
18	25	3.25
	50	3.04
	100	4.42

[a] Geometric mean anti-A concentration determined by radioactive antigen binding assay

Dextran 200, and Vi is the total volume of the liquid in the gel bed measured with (^{14}C) sodium acetate[21].

The standard molecular size recommended by The Division of Biologics, FDA, is that at least 50% of the polysaccharide must be eluted with a Kd of 0.5 or smaller[21]. The WHO Expert Committee on Biological Standardization recommends that at least 65% of the group A polysaccharide (and 75% of the C polysaccharide) be eluted before a Kd of 0.5 is obtained[21].

Gotschlich, *et al*. showed that lots of group A vaccine with Kd values of 0.5 and 0.6 respectively were significantly less immunogenic than lots with Kd values of 0.37 and 0.27[25]. The dependence of immunogenicity on molecular size has been confirmed in infants[12] and is of critical importance in the protective efficacy of the vaccine. Gold, et al. showed that although 2 doses of Lot A-5 (Kd 0.37) given at 3 and 12 months of age induced geometric mean anti-A concentrations of 4 μg/ml, the peak anti-A concentration after 2 doses of lot A-7 (Kd 0.57) was only 1 μg/ml. Only 26% of infants had 2.0 μg/ml after lot A-7 compared to 79% after Lot A-5 administered at 3 and 12 months of age[12,17].

Interval Between Primary and Booster Immunization. No significant differences in antibody responses were seen when infants were immunized at 3 and 7 months of age[12] compared to 3 and 6 months of age[1]. Whether the interval between the 2 doses of vaccine can be reduced to 2 months or less is currently under investigation. Three doses of vaccine administered at 2, 4, and 6 months of age resulted in significantly lower anti-A concentrations than the previously used 2 dose schedule[15]. Whether the reduced response was the result of the shorter interval between immunizations or to smaller molecular size of the vaccines used in the 3 dose compared to the 2 dose trials remains to be determined.

Persistence of Antibody. After primary immunization of 20 children 2-11 years of age, anti-A declined by 50% from a peak value of 8.77 μg/ml to 4.45 μg/ml within a 2 year period[16]. The children were given a booster 3 years after the primary and followed for an additional three years. Mean anti-A level declined 32% one year after the booster, but showed no significant further decrease over the next two years (Table 9). A larger study of school children 6-8 years of age showed similar results: namely,

TABLE 9

PERSISTENCE OF ANTI-A ANTIBODY FOLLOWING PRIMARY AND BOOSTER IMMUNIZATION OF CHILDREN 2-11 YEARS OF AGE

Time after 1° & 2° Immunization		Anti-A Concentration (µg/ml)[a]	%>2.0 µg/ml[b]
Primary	Pre	1.30	37
	1mo	8.77	100
	1yr	6.84	100
	2yr	4.45	94
Booster[c]	1mo	13.08	100
	1yr	8.71	100
	2yr	7.97	100
	3yr	7.85	100

a Geometric mean anti-A concentration determined by radio-immuno assay.

b % of children with ≥2.0 µg/ml of anti-A antibody 1 month after immunization.

c Children were given a booster (2°) injection of vaccine 3 years after primary (1°) immunization; serum samples were assayed for antibody at the time intervals shown.

4 years after immunization, the anti-A concentration was 50% of the peak level (Table 10). However, 80% of the children still had ≥2 µg/ml of anti-A antibody.

Persistence of antibody after immunization of infants is relatively short. Anti-A concentrations return to pre-immunization base line values 12-18 months after the booster dose at 6-7 months of age[1,12,14,17]. Booster immunization at 2 years of age results in anti-A concentrations similar to that seen in older children[14].

TABLE 10

PERSISTENCE OF ANTI-A ANTIBODY IN SCHOOL CHILDREN AFTER PRIMARY IMMUNIZATION OF CHILDREN 6-8 YEARS OF AGE WITH GROUP A MENINGOCOCCAL POLYSACCHARIDE VACCINE

Time after Immunization	Anti-A Concentration ($\mu g/ml$)[a]	% with > 2.0 $\mu g/ml$[b]
0	1.68	61
1mo	9.35	98
1yr	5.54	89
3yr	4.38	90
4yr	3.62	80

[a] Geometric mean anti-A concentration determined by radio-immuno assay.

[b] % of children with > 2.0 $\mu g/ml$ of anti-A antibody 1 month after immunization.

Preliminary results indicate that the anti-A levels declined to that of unimmunized children by 5-6 years of age [17]. Thus, maintenance of anti-A at concentrations ≥2.0 $\mu g/ml$ during infancy and childhood will require 2 doses of vaccine in infancy, followed by booster injections at 18 months and 5 years of age.

Group C Meningococcal Polysaccharide

Age. In contrast to their lack of response to the group A vaccine at 3 months of age, over 90% of infants show a significant increase in anti-C antibody following primary immunization with the group C polysaccharide[12]. With increasing age, there is a linear increase in peak anti-C concentration (Table 11). The increasing response to the C vaccine with age is also apparent in the proportion of children with ≥ 2.0 $\mu g/ml$ of anti-C antibody one month after primary immunization. By 18 months of age, 72% of children achieve ≥ 2.0 $\mu g/ml$.

TABLE 11

ANTIBODY RESPONSES OF INFANTS AND CHILDREN TO PRIMARY IMMUNIZATION WITH GROUP C MENINGOCOCCAL POLYSACCHARIDE VACCINE

	Anti-C Concentration (μg/ml)				
	Pre[a]		Post[b]		
Age	N	G.M.[c]	N	G.M.	%>2.0 μg/ml[d]
3mo	85	0.11	79	0.37	3
7mo	69	0.11	75	0.97	20
12mo	38	0.12	38	2.10	58
18mo	60	0.27	55	3.71	72
2-5yr	34	0.18	33	5.54	88
6-8yr	90	0.33	125	9.12	95

[a] Pre = pre-immunization level of serum antibody.

[b] Post = antibody level 1 month after immunization.

[c] G.M. = geometric mean anti-C concentration (μg/ml of serum) for N individuals, determined by radioimmunoassay.

[d] Percent of children with > 2.0 μg/ml of anti-C antibody 1 month after immunization.

Prior Experience With Group C Polysaccharide. Natural immunization to the group C polysaccharide occurs at a much slower rate than to the group A polysaccharide. At 1 year of age, only 25% of infants have detectable anti-C antibody compared to 67% with anti-A antibody[12,17]. At 6-8 years of age, the geometric mean anti-C concentration is 0.34 μg/ml and only 7% of children have ≥2.0 μg/ml in anti-C[16]. In contrast the geometric mean anti-A concentration at 6-8 years of age is 1.68 μg/ml and 61% of children have $\geq$2.0 μg/ml of anti-A antibody. Nevertheless, despite the lack of prior natural exposure to the group C polysaccharide, infants and children respond to primary immunization with the group C polysaccharide just as well as they do to the group A polysaccharide in terms of peak antibody concentrations. Indeed, unlike the group A polysaccharide, no correlation was observed between pre-immunization anti-C concentration and the response to primary immunization[12,17].

An even more striking difference in the responses of infants to the groups A and C polysaccharides was observed after booster immunization[12]. In contrast to the significant anamnestic responses seen after booster doses at 7 or 12 months of age with the group A vaccine, booster immunization with 25 or 100 μg of group C polysaccharide at 7 or 12 months of age resulted in significantly lower anti-C concentrations compared to primary immunization at these ages (Table 12). The suppression of the anti-C response varied with age, dose of C vaccine, and interval between immunizations. When infants were given 10 μg of group C polysaccharide at 3 and 7 months of age, the peak geometric mean anti-C concentration was the same as that induced by primary immunization at 7 months. Moreover, when infants received their first dose of C vaccine at 7 months of age, no suppression was seen after a booster at 12 months of age. No hyporesponsiveness was noted in children 2-10 years of age given 2 doses of group C polysaccharide 2 months apart[17]. However, hyporesponsiveness did occur in U.S. Army recruits given 2 doses two weeks apart[32]. The recruits were immunized with Lot A-7 of group A polysaccharide which contained trace amounts of group C polysaccharide (approximately 0.03 μg of

TABLE 12

ANTIBODY RESPONSES OF INFANTS TO PRIMARY AND BOOSTER IMMUNIZATION WITH GROUP C POLYSACCHARIDE VACCINE

Age(mo)	Vaccine Dose(μg)	Vaccine Schedule[a]	Anti-C Concentration(μg/ml)[b]	%>2.0 μg/ml[c]
7	10	1° at 7mo	0.64	1
		1° at 3mo and 2° at 7mo	0.77	0
	25	1° at 7mo	0.80	14
		1° at 3mo and 2° at 7mo	0.23	0
	100	1° at 7mo	1.41	27
		1° at 3mo and 2° at 7mo	0.32	5
12	10	1° at 12mo	1.08	15
		1° at 7mo and 2° at 12mo	1.04	29
	25	1° at 12mo	1.69	21
		1° at 3mo and 2° at 12mo	0.94	39
	100	1° at 12mo	2.62	68
		1° at 3mo and 2° at 12mo	0.76	17
		1° at 7mo and 2° at 12mo	3.02	57

a 1° (primary) or 2° (booster) immunization given at age indicated.

b Geometric mean anti-C concentration determined by radioimmuno assay.

c % of children with > 2.0 μg/ml anti-C antibody 1 month after immunization.

C per dose of A). The antibody responses of these recruits to the group C polysaccharide administered 2 weeks later were significantly lower than those of recruits who received only the group C vaccine. This same lot of group A polysaccharide did not induce hyporesponsiveness to group A vaccine in infants when it was administered at least 4 months before the group C polysaccharide[12,17].

The hyporesponsiveness induced by primary immunization at 3 months of age disappears by 2 years of age. Infants who had received 1 or 2 doses of group C polysaccharide between 3 and 12 months of age had normal antibody responses to booster immunization at 2 years of age[14,17].

The factors responsible for the hyporesponsiveness observed after immunization of infants with group C polysaccharide are unknown. The lack of immunogenicity of the purified meningococcal polysaccharides in animals[33] make studies of the roles of the various T cell subpopulations and their interactions with B cells difficult to undertake.

Dose of Group C Polysaccharide. The antibody responses of infants to primary immunization with the group C meningococcal polysaccharide are dose-dependent. At 3, 7, and 12 months of age, significant differences in the peak anti-C concentrations were observed after 10, 25 or 100 μg of group C polysaccharide (Table 13)[12,17]. Similar dose-related differences were reported in adults, with 10 μg of group C polysaccharide inducing less antibody than 50 μg[29].

Molecular Size of Group C Polysaccharide. Most studies reported to date on the immunogenicity of group C polysaccharide in infants have used Lot C-10 which has a Kd of 0.3. The molecular size of the vaccine used in Sao Paulo, Brazil has not been reported, but the antibody responses of the infants receiving that

TABLE 13

ANTIBODY RESPONSES OF INFANTS TO PRIMARY IMMUNIZATION WITH DIFFERENT DOSES OF GROUP C MENINGOCOCCAL POLYSACCHARIDE VACCINE

Age (mo)	Dose (µg)	Anti-C Concentration (µg/ml) [a]
3	5	0.17
	10	0.25
	25	0.43
	100	0.49
	200	0.28
7	10	0.73
	25	0.80
	100	1.41
12	10	1.08
	25	1.69
	100	2.62
18	25	2.98
	50	3.09
	100	5.73

[a] Geometric mean anti-C antibody concentration determined by radioimmuno assay.

vaccine were significantly lower than those of American infants of the same age unimmunized with Lot C-10[12,34]. Recently, lots of group C polysaccharide which are significantly larger in molecular size have become available[13]. Although no significant differences were observed in the immunogenicity of different lots of C vaccine in children 2-10 years of age, the larger group C polysaccharide with Kd 0.24 induced significantly higher anti-C antibody concentrations in infants 3 and 7 months of age[17]. Further studies on the relationship of molecular size to immunogenicity are in progress and are of great importance because of the lack of protection afforded infants under 2 years of age with standard group C vaccine[6].

Persistence of Antibody. The rate of decline of anti-C antibody concentration following primary immunization is much more rapid than that occurring after immunization with group A vaccine[12,16]. One year after immunization of children 6-8 years of age, the geometric mean anti-C concentration was only one-third of the peak value and only 55% of children had $\geq$2.0 μg/ml[16]. By 4 years after immunization, the anti-C concentration was 1/4 of the peak value and only 40% of children had $\geq$2.0 μg/ml (Table 14).

The decline of anti-C antibody is even more rapid in infants. By 2 years of age, the geometric mean anti-C concentration of infants previously immunized was 0.2-0.4 μg/ml, only slightly greater than that of unimmunized infants[12,14]. Following a booster at 2 years of age, peak anti-C concentrations of 2-4 μg/ml were induced[14]. By 3 years of age, the anti-C level declined to 1 μg/ml, and by 5 years of age, the geometric mean anti-C concentration was 0.7 μg/ml, only slightly greater than that of unimmunized children[17]. The rapid decline of antibody following immunization of infants and young children is a major obstacle to

TABLE 14

PERSISTENCE OF ANTI-C ANTIBODY FOLLOWING PRIMARY IMMUNIZATION AT 6-8 YEARS OF AGE

Time After Immunization	Anti-C Concentration (μg/ml)[a]	% > 2.0 μg/ml[b]
0	0.33	7
1mo	9.12	95
1yr	2.35	55
3yr	1.79	48
4yr	1.47	40

[a] Geometric mean anti-C concentration determined by radioimmuno assay.

[b] % of children with > 2.0 μg/ml of anti-C antibody 1 month after immunization.

routine use of the presently available vaccines.

Combined Group A and C Polysaccharide Vaccines. The responses to the A and C polysaccharides are group-specific. No significant differences were observed in the responses of infants to administration of groups A and C polysaccharide simultaneously in separate sites compared to bivalent A and C vaccine given in one site Table 15)[14,17]. Similar findings have been reported in adults[35,36].

Group B Meningococcal Polysaccharide

Purified group B polysaccharide vaccines have been prepared by methods similar to those used for the groups A and C polysaccharides[37]. However, the purified polysaccharides failed to induce any anti-B antibody in adult volunteers[37,38]. The lack of immunogenicity of the purified polysaccharide is in contrast to the brisk anti-B polysaccharide antibody responses seen after group B meningococcal disease. The marked acid-lability and sensitivity

TABLE 15

ANTIBODY REPONSES OF INFANTS TO MONOVALENT AND BIVALENT MENINGOCOCCAL POLYSACCHARIDE VACCINES

Age(mo)	Vaccine[a]	Antibody Concentration (μg/ml)[b]	
		Anti-A	Anti-C
24	A & C Separate Sites	6.48	3.00
	A & C Combined	5.29	3.34

[a] Separate sites = A and C administered simultaneously in separate sites; Combined = bivalent A and C vaccine administered in single site.

[b] Geometric mean anti-A or anti-C concentration determined by radioimmuno assay.

to neuraminidase of the group B polysaccharide may contribute to the failure of the vaccines.

An alternative approach to an effective vaccine for group B meningococcal disease has been to use the serotype proteins of the outer membrane. Serotype 2 protein vaccines have been shown to induce protective antibodies in a guinea pig model of meningococcal infection[39] and in the mouse-mucin model[40]. The vaccines also induced significant increases in bactericidal antibody against group B and group C strains in adult volunteers[40]. Investigation of such protein antigens and also of means of enhancing the immunogenicity of the group B polysaccharide should receive continued support. The lack of an effective group B vaccine remains a major unsolved problem in the control of meningococcal disease.

<u>Group Y Meningococcal Polysaccharide</u>

A group Y meningococcal polysaccharide vaccine has been prepared and evaluated in adults[41]. No significant adverse reactions were observed. Significant increases in bactericidal antibody titers were induced in 93% of the ninety adult volunteers.

EFFICACY OF GROUPS A AND C MENINGOCOCCAL POLYSACCHARIDES

The group A meningococcal polysaccharide vaccine has been shown to be highly effective in preventing disease in infants, children, and adults. During the epidemic of group A meningococcal disease in Finland, the vaccine was 90% effective among military recruits[5] and 100% effective in infants 3 months to 5 years of age[1]. Field trials of the vaccine in Egypt also showed over 90% efficacy in preventing disease among school children 6-15 years of age[2,4]. Duration of protection was 2 years in the Egyptian studies.

The group C meningococcal polysaccharide vaccine has also been highly effective in preventing disease in adults. Field trials in the U.S. Army demonstrated an 87% reduction in the incidence of group C meningococcal disease as a result of immunization[7,8]. Moreover, the group C vaccine also reduced the rate of acquisition of the carrier state[7,29]. However, no protection was observed in infants 6-23 months of age during field trials of the group C vaccine in Brazil [6]. The vaccine reduced the incidence of disease by 75% in children 24-35 months of age.

CONCLUSIONS

The groups A and C meningococcal polysaccharide vaccines have been found to be safe, immunogenic and effective in preventing disease. Control of epidemic group A meningococcal disease among all age groups should be feasible with currently available vaccines. Moreover, prevention of group A disease by routine immunization of children may also be achieved. Immunization at 3 and 6 months of age, followed by boosters at 1 1/2 and 5 years of age would appear to provide protective levels of anti-A antibody throughout childhood.

Two currently available group C vaccines appear to be sufficiently immunogenic to induce protective antibody levels in children 18 months and older. A single dose of group C vaccine for children over 18 months of age should be effective in the event of an epidemic of group C disease. Protection of younger infants may be possible with the larger molecular weight vaccines. In addition, the reduction in transmission of group C strains by the effect of the vaccine on the acquisition of the carrier state may also provide indirect protection to infants. Because of the relatively short persistence of anti-C antibody in young children, routine immunization with currently available group C vaccine cannot be recommended.

Control of epidemic and endemic meningococcal disease will not be achieved until several major problems are solved. The immunogenicity of the group C vaccine in infants must be improved. Whether increasing the molecular size of the group C polysaccharide is the answer remains to be determined. Even more important is the lack of a vaccine against the group B meningococcus. Evaluation of serotype protein vaccines has begun, but the relative importance of the different group B serotypes in causing disease and the safety and immunogenicity of such vaccines remains to be determined.

A group Y polysaccharide has been developed and been found to be safe and immunogenic in adults. The need for such a vaccine as as well as vaccines against the less frequent meningococcal serogroups X, Z, 29E, and W135 is unknown.

ACKNOWLEDGEMENT

The studies were performed in collaboration with Drs. Martha L. Lepow and Irving Goldschneider of The University of Connecticut, Dr. Emil C. Gotschlich of The Rockefeller University, Dr. Thomas Draper of The Danbury Connecticut Health Department, and Dr. Martin Randolph of Danbury, Ct. Research support was provided in part by Research Contracts No. N01-AI-22502 and N01-AI-12502 from The National Institute of Allergy and Infectious Diseases, by Contract No. DAPH-17-70-C-0027 from the U.S. Army Research and Development Command, by Research Grants from Merrell-National Laboratories, and by Research Grant CC0648 from The U.S. Public Health Service.

REFERENCES

1. Peltola, H., Makela, P.H., Kayhty, J., Jousimies, H., Herva, E., Hallstrom, H., Sivonen, A., Renkonen, O.-V., Pettay, O., Karanko, V., Ahvonen, P. and Sarna, S. (1977) N. Eng. J. Med. 297, 686-691.

2. Wahdan, M.H., Rizk, F., El-Akkad, A.M., El Ghoroury, A.A., Hablas, R., Girgis, N.I., Amer, A., Boctar, W., Sippel, J.E., Gotschlich, E.C., Triau, R., Sanborn, W.R. and Cvjetanovic, B. (1973) Bull. WHO, 48, 667-673.

3. Erwa, H.H., Haseeb, M.A., Idris, A.A., Lapeyssonie, L., Sanborn, W.R. and Sippel, J.E. (1973) Bull. WHO, 49, 301-305.

4. Wahdan, M.H., Sallam, S.A., Hassan, M.N., Gawad, A.A., Rakha, A.S., Sippel, J.E., Hablas, R., Sanborn, W.R., Kassem, N.M., Riad, S.M. and Cvjetanovic, B. (1977) Bull. WHO, 55, 645-651.

5. Makela, P.H., Kayhty, H., Weckstrom, P., Sivonen, A. and Renkonen, O.-V. (1975) Lancet, 2, 883-886.

6. Taunay, A. de E., Galvao, P.A., deMorais, J.A., Gotschlich, E.C. and Feldman, R.A. (1974) Ped. Res. 8, 429.

7. Artenstein, M.S., Gold, R., Zimmerly, J.G., Wyle, F., Schneider, H. and Harkin, C. (1970) New Eng. J. Med. 282, 417-420.

8. Gold, R. and Artenstein, M.S. (1971) Bull. WHO, 45, 279-282.

9. Artenstein, M.S., Winter, P.E., Gold, R. and Smith, C.D. (1974) Mil. Med. 139, 91-95.

10. Goldschneider, I., Lepow, M.L. and Gotschlich, E.C. (1972) J. Infect. Dis. 125, 509-519.

11. Goldschneider, I., Lepow, M.L., Gotschlich, E.C., Mauck, F.T., Bachl, F. and Randolph, M. (1973) J. Infect. Dis. 128, 769-776.

12. Gold, R., Lepow, M.L., Goldschneider, I., Draper, T.F. and Gotschlich, E.C. (1976) J. Clin. Invest. 56, 1536-1547.

13. Gold, R., Lepow, M.L., Goldschneider, I., Gotschlich, E.C., DeSanctis, A.N. and Metzgar, D.P. (1977) Pediat. Res. 11, 500.

14. Gold, R., Lepow, M.L., Goldschneider, I. and Gotschlich, E.C. (1977) J. Infect. Dis. 136, S31-S35.

15. Gold, R., Lepow, M.L., Goldschneider, I. and Gotschlich, E.C. (1978) J. Infect. Dis., In press.

16. Lepow, M.L., Goldschneider, I., Gold, R., Randolph, M. and Gotschlich, E.C. (1977) Pediatrics, 60, 673-680.

17. Gold R., Lepow, M.L., Goldschneider, I. and Gotschlich, E.C. Unpublished data.

18. Apicella, M.A. (1974) J. Infect. Dis. 129, 147-153.

19. Gotschlich, E.C. (1975) Monogr. Aller. 9, 245-248.

20. Gotschlich, E.C., Liu, T.Y. and Artenstein, M.S. (1969) J. Exp. Med. 129, 1349-1365.

21. Wong, H.H., Barrera, O., Sutton, A., May, J., Hochstein, D.H., Robbins, J.D., Robbins, J.B., Parkman, P.D. and Seligman, E.B., Jr. (1977) J. Biol. Stand. 5, 197-215.

22. WHO Expert Committee on Biological Standardization (1976), WHO Tech. Rep. Series, 594, 19-75.

23. WHO Expert Committee on Biological Standardization (1977), WHO Tech. Rep. Series, 610, 52-53.

24. Artenstein, M.S., Brandt, B.L., Tramont, E.C., Branche, W.C., Jr., Fleet, H.D. and Cohen, R.L. (1971) J. Infect. Dis. 124, 277-288.

25. Gotschlich, E.C., Rey, M., Triau, R. and Sparks, K.J. (1972) J. Clin. Invest. 51, 89-96.

26. Goldschneider, I., Gotschlich, E.C. and Artenstein, M.S. (1969) J. Exp. Med. 129, 1327-1348.

27. Gold, R., Lepow, M.L., Goldschneider, I. and Gotschlich, E.C. (1978) J. Infect. Dis., In press.

28. Gold, R., Lepow, M.L. and Randolph, M. (1978) Ped. Res. 12, 492.

29. Artenstein, M.S., Gold, R., Zimmerly, J.G., Wyle, F.A., Branche, W.C., Jr. and Harkins, C. (1970) J. Infect. Dis. 121, 372-377.

30. Peltola, H., Kayhty, H., Kuronen, T., Haque, N., Sarna, S. and Makela, P.H. (1978) J. Ped. 92, 818-822.

31. Farquhar, J.D., Hankins, W.A., DeSanctis, A.N., DeMeio, J.L. and Metzgar, D.P. (1977) Proc. Soc. Exp. Biol. Med. 155, 433-455.

32. Artenstein, M.S. and Brandt, B.L. (1975) J. Immunol. 115, 5-7.

33. Gotschlich, E.C., Goldschneider, I. and Artenstein, M.S. (1969) J. Exp. Med. 129, 1367-1384.

34. Amato Neto, V., Finger, H. Gotschlich, E.C., Feldman, R.A., Avila, C.A. de, Konicki, S.R. and Laus, W.C. (1974) Rev. Inst. Med. Trop. Sao Paulo, 16, 149-153.

35. Weibel, R.E., Villarejos, V.M., Vella, P.P., Woodhour, A.F., McLean, A.A. and Hilleman, M.R. (1976) Proc. Soc. Exp. Biol. Med. 153, 436-440.

36. Brandt, B.L., Smith, C.D. and Artenstein, M.S. (1978) J. Infect. Dis. 137, 202-205.

37. Wyle, F.A., Artenstein, M.S., Brandt, B.L., Tramont, E.C., Kasper, D.L., Altieri, P.L., Berman, S.L. and Lowenthal, L.P. (1972) J. Infect. Dis. 126, 514-522.

38. Kasper, D.L., Winkelhake, J.L., Zollinger, W.D., Brandt, B.L. and Artenstein, M.S. (1973) J. Immunol. 110, 262-268.

39. Frasch, C.E. and Robbins, J.D. (1978) J. Exp. Med. 147, 629-644.

40. Zollinger, W.D., Mandrell, R.E., Altieri, P., Berman, S., Lowenthal, J. and Artenstein, M.S. (1978) J. Infect. Dis. 137, 728-739.

41. Farquahar, J.D., Hankins, W.A., DeSanctis, A., DeMeio, J.L. and Metzgar, D.P. (1977) Proc. Soc. Exp. Biol. Med. 155, 453-455.

Index

notes